SOL

CENTURION

By
E. Clifton Sosebee

1

To

Sarah Elizabeth Sosebee
****

**My Life, My Heart, My Soul,
and God's physical and spiritual
manifestation testimony for myself,
that truly reveals The Almighty
Spirit's utmost power.
His greatest reward to his children,
known as mankind.**

His Gift of Love.

Author's Note

This little novella will always be my pride and joy. I wrote it years ago, to the rhythm of several songs.

I got the idea when I saw the second part of a three part interview of George Lucas by Leonard Maltin.

One night in the late 80's, I learned there was going to be a lunar eclipse that night. After sunset, I jumped into my ole' 73, cushy ride Buick, and headed away from the city's overwhelming lights to get a better view of that mildly rare celestial event.

Just as the eclipse began, one of the local Rock stations began playing the entire album, "The Dark Side Of The Moon," by Pink Floyd to coincide with the eclipse. When the third song on the album began, my imagination began drifting with the increasing rhythm of the power from that song, and I somehow got an image of a small alien spacecraft swiftly coming into our Solar System just outside the orbit of Pluto.

After perhaps a minute into that song, I then heard and imaged from the intensifying powerful force of that magnificent music, an incredible, huge tremendous alien vessel with a much more meaner, malicious look about it, leaping out from somewhere in the Kuiper Belt at tremendous speed and swinging directly into the little ship's wake, then rapidly increasing its velocity by each passing moment to overtake it.

Several times within the instrumental might of that song you'll hear the smaller ship pass you now heading for cover, followed by a few moments later by the trailing massive predator passing you swiftly closing in for the kill. Just before the song comes to its end, you'll hear an enormous rapid upsurge of powerful energy release, instantaneously followed by a titanic destructive detonation.

I saw that sequence as the immense alien spacecraft's shadow slowly sweeping over and engulfing the fleeing smaller one below, that was now soaring and shifting erratically every which way it could in one last attempt to escape their inevitable fate by this incredibly huge, horrifying menace by quickly re-altering its inconsistent course over the mountains and valleys of the planet Mars.

The unfortunate little vessel tried everything it could to somehow evade this unbelievable terrifying atrocity now directly above them, but in one last moment to grasp at survival, was effortlessly annihilated in one dynamic burst of radiating spear of light.

Perhaps six or seven seconds after the explosion, you can hear a far distant echo of what's left of the little ship's cremated fuselage impact onto the wall of one of the towering mountain sides, majestically spread out upon the red Maritain surface.

I was pretty much awestruck at what my imagination illustrated back to me in that three and a half minute musical sequence, and thought, "Wow. So much for the poor saps on that little ship. That big bad ship really needs to pick on something its own size. Bully. Hate when that happens to me."

Suddenly the next track kicked in, which is a bunch of blasting alarm clocks going off all at once, and almost demolished my cushy glide ride Buick into the side of an eighteen wheeler while spastically reaching out to turn the volume down that naturally my right hand couldn't find, while trying my best to hold my swaying Buick on the road.

Later that evening I kept replaying that little chase scenario in my mind, and thought that could be the beginning of a good Sci-Fi story. A small ship, or at least small compared to the huge one on safari, coming into our system, for what? Something on the Earth? Okay. But what? Something on the Earth, that perhaps a superior alien race has discovered and is now protecting it? Alright, that could work. Especially with why that far superior alien ship blew the little one back into the Stone Age.

Went to a record store the next day and bought, "The Dark Side Of The Moon," CD. Played it when I got home just to see what the name of that song was, knowing it would be the one just before those insane alarm clocks went off. It blew me away when I saw what the name was. "On The Run," and thought that's exactly what I saw.

The thought kept returning in the weeks that followed as to why a far superior alien civilization would protect the Earth from all the other ones. A treasure of some kind, that exists only on the Earth. What would really be cool if it had something to do with human beings? Yeah, right. A race of beings that slaughters each other for

4

religion, land and greed. However, there must be something about us that's special. What special mystery about us could they be protecting us from all the other races from?

A few nights later something sat me straight up in bed. Human beings only use ten to fifteen percent of their brain cells. Carl Sagan once said that he believed the greatest force in the known universe is not electrical, gravitational nor even nuclear...it's the power of human thought. Bingo! Thank you Carl!

Okay, I will now ask.

To all the world's scientists, and to all those who have ever took even a few minutes of time to ever wonder what would actually happen if the eighty-five to ninety percent of our unused, dormant brain cells were somehow, someway brought to life?

I truly believe, in all reality, that Carl Sagan's hammer struck that nail's top straight away top dead center perfectly right on target. He is absolutely, positively correct.

The possibilities of a fully utilized human brain could be so vastly enormous that it not only could, but would actually be too inconceivably incomprehensible for even Einstein himself to possibly dare to even think of what might happen.

This little fictional tale is based on one possible theory of what might just result if only a portion of those inactive cells were someway turned on and became fully functional.

Yet, the real question in this story still remains an unanswered mystery.

Why?

Why are those cells there, but only at rest? It's very simple math. There is an answer, because they wouldn't be there if there wasn't.

One of my greatest loves will always be the incredible mysteries of physics. I did not wish to contend with the possible theories of the answer to that one in this story.

But I do intend to make Sol Centurion into a trilogy, and will come up with some type of answer as to why they exist but are not utilized in the present stage of human evolution.

So enjoy this one for now, and do not think when you read the last page that this little story is over. Because when you do, all I can say for now is the story at that point is only the beginning.

CHAPTER 1

ON THE RUN

The greatest treasure in the universe.

Who would be so bold as to even conceive of attempting to achieve its possession? To capture such power would grant any civilization, of any world in the universe the power to do as they wish, without resistance of any nature or form.

They would. They must.

In every race, lives those who must reach for the most ultimate trophy of existence, to satisfy their sense of accomplishment, no matter what the cost.

But the cost. The greatest reward could only be obtained at the greatest price. Many have attempted, but all have failed. Any who have tried are never heard from again.

The immense reward that exists on the third planet of the approaching system could be taken with no opposition from the indigenous inhabitants. But what has stopped any attempt to get close to this world, lurks nearby, only to reveal itself once a trespasser comes within the boundary orbits of the outer gas giant planets of the system.

Once this threshold is crossed, they simply vanish from existence.

But with the greatest fortune of the universe at stake, who could resist? They must try.

Passing the outer cloud of asteroids that encircles this system, they still proceeded inward. There was no sign of whatever protects this little blue-white planet, but it was well known they would have to contend with it at some point.

Yet unknown to them, still a good distance away, a massive, magnificent sentinel was sliding into their wake at astonishing speed.

They crossed the orbit of the outermost planet, knowing now they were fully committed.

No sign of anything yet.

They slowly approached the giant ringed planet, and hid within its shadow. Still no signal of any movement within their proximity that would indicate danger, but still unaware the chase had already begun.

After waiting long enough, they moved with extreme caution out of the ringed planet's massive shadow. Once clear of the exquisite planet's outer ring, they adjusted their course to stay in the shadow but bypass the star system's largest planet, the mighty multi-colored banded gas giant, and head for the fourth planet, a small rusty red world. There they could hide again in the shadow of this planet, to wait and see if there was any indication of any kind of pursuit.

Just as they closed in upon the belt of asteroids between the star system's largest gas giant and the little red world, from out of nowhere it came, closing on them fast.

They swiftly maneuvered between the shifting and swaying space tumbling rocks, and dashed for the red world with their utmost speed.

What they detected pursuing them was some sort of vessel, of enormous size. It was parting the asteroids when it entered the belt of asteroids, as if the huge rocks were commanded to step aside, still bearing down on them at incredible speed.

Bursting through the atmosphere of the red planet, they quickly headed for the surface, hoping to hide somehow within the great mountains and valleys that covered this desert world.

The gap between them and this massive predator was narrowing fast. They slipped over the highest mountain with great haste, but the gigantic marauder on their trail shadowed the entire mountain, as if it were only a hill of dust.

No matter which way they shifted or turned, the silhouette of their incredible stalker was beginning to overtake them.

A valley was ahead. Heading downward into the valley, they soared just above the valley's bottom surface, slipping in and out of the canyons and crevasses submerged within the depths of this valley's vast enormity.

Maneuvering with all possible speed, they saw the edge of the valley coming to an end. They charged up the valley wall, trying to reach the top, but their pursuer's shadow now completely engulfed the entire valley.

Swiftly jumping over the valley's edge, they hugged to the surface as close as possible. In one last desperate effort, they headed toward the next mountain range.

In one astonishing majestic motion, an array of blazing rich colors flared out from both ends of the immense hunter, instantaneously rolled from each end to collide at the ship's center, which produced a thin blinding streak of light, targeting the fleeing ship.

The escaping ship exploded into thousands of remains, scattering and rolling fragments far and wide over the desert of the little red world.

They, like the rest, would never bear witness to anyone else of what they encountered during their quest for the greatest treasure in the universe.

The shadow over the great valley of the red planet turned and slowly faded away, returning to wherever it lurks within this system, to await the return of the next, cursed with the obsession of owning the greatest wealth in existence.

The third little blue-white planet remained intact, untouched and unharmed. But first and foremost, totally unaware of the greatest jewel in the universe placed upon its land.

CHAPTER 2

DEPARTURE

The day finally came. His dream would now be fulfilled at last. All that stood in his way before was now behind him.

Stepping through the hatchway, he turned for the last time to look out at the world he wished to never see again, then turned and sealed the hatch shut.

Strapping himself in, he reached up and switched on the power controls. He leaned back in his pilot's seat and listened to the roar of the engines gradually begin to increase, and the little cabin started to tremble.

Once the instruments flashed that the reactors were fully generated, he engaged the throttle controls. He could feel the ship ease itself from the ground, and realized his journey had finally begun.

The ship was finally high enough, so he adjusted the attitude control to put the ship in an almost vertical position, and eased the main throttle forward. He held on as the little ship streaked toward the upper atmosphere.

As his velocity increased, he could feel the mighty hand of the Earth's gravity gradually pulling him back into his seat, and he thought of all the troubles and problems he had dealt with to achieve his dream. This was the last strain the Earth could burden him with.

It was not until the stars burst out on the forward viewing screen that the G forces slowly lifted their fingers from him and his little one-manned ship, and at last he was totally free from all the Earth's influence.

He was on a journey that countless men dreamed of, but for David Adams, it was now reality. He walked to the rear of the ship, and through the stern viewing port David watched the bluish-white world descending away into the blackness of space, thinking of all who laughed and rejected all of his theories of how to finally grasp and harness the power of not just nuclear fission, but the same incredible power that engineers the sun itself; the magnificent might of nuclear fusion, that would have revolutionized the evolution of mankind.

"Nonsense, they said," he thought to himself, "And look at them now, all gathered together in a world of mental stagnation, where they belong."

As he turned and walked to the forward viewing port, he could feel his bitterness for his former colleagues slowly increasing. As he looked out of the forward screen, he watched the beauty of the never-ending universe expanding before him.

His past. With the rapture of this vast ocean of eternal magnificence unfolding before him with each passing moment, his burning resentment began to drift back further. Back to the time as a child that was avoided and laughed at, not only by his friends, but even his own family, because of his obsessive interest in the stars and planets, and what inconceivable wonders lay beyond them. The hurt still remained within, and could never understand throughout his adolescence, teenage years and college, why his greatest passion to study the universe, made him an outcast of some kind; even by his short marriage as well.

He sat and leaned back in his chair, wondering with excitement about his future destiny. For hours he stared out, looking at the billions of stars that lay in his path, when finally the past long days of enduring turmoil got the best of him, and he eased off to sleep. He dreamed of the worlds he was to encounter, of empires and alien civilizations welcoming him as if he was a representative sent by a highly intelligent race, gathering information and culture, then moving on to the next world. Each new world David encountered he found even more fascinating and beautiful than the last.

Floating through fantastic rivers of gas clouds inside great nebulas, seeing mighty pulsars throbbing their nuclear winds out into the depths of space, like celestial lighthouses beaconing throughout the universe, he witnessed an eternity of incredible beauty and wonder. He knew then there was no end to the glory of God's creations.

Drifting through his dreams he could feel something was not right. At first he could not pinpoint it. As he awoke, he suddenly realized the automatic sensor alarm was on. Switching off the alarm, he engaged the sensor screen, and turned to the navigation instruments. They showed his position just beyond the orbit of Jupiter. Turning back to the sensor screen, David could not believe what he was seeing.

The sensor showed an enormous object approaching, miles long, coming up behind him, closing fast. He ran to the rear viewing port and looked. "What am I looking at?" he thought, paralyzed by a state of shock that overcame him. Bright glistening lights, almost as bright as the sun itself, were quickly filling the viewing screen, blinding him as the object approached.

Shielding his eyes with his hands and slowly walking backwards, he began to tremble. He started shaking more with each step, until bumping the back of his pilot's chair. He slowly slid down the back of his chair, then crumpled himself into a ball, shaking and horrified beyond any capability of rational thought. The interior of his ship became so bright with the blazing radiance that visibility completely vanished, as he slipped into unconsciousness.

CHAPTER 3

ENCOUNTER

"So this is death," he thought, as lights began peering into his awareness. They were lights with such brilliance and colors, which actually seemed to touch him, and gave him a beautiful calmness he had never before felt, yet he was unable to move.

Suddenly, he remembered what had happened. He was on his little ship heading into the unknown, when the ship was abruptly overtaken by some kind of strange lights, which he was seeing now. David began to feel himself breathing. Slow, short breaths, that got deeper with each mouthful of air.

Still unable to move, he concentrated on the lights, which seemed like thousands of different colors, each with its own separate design and shape, dancing rapidly between each other. The colors began to gradually fade into a hazy, glistening white, yet safe now to look at. Within himself, he began to feel control of his movements, so he rose up slowly. He appeared to be in some kind of small room, with that same white gleaming haze surrounding the entire area, but with no door.

"So there is life after death," David thought again, "but I didn't think my body would come with it!" He slowly rose to his feet, walked to one side of the room to what appeared to be one of the walls, with the cloudy haze just floating above the wall surface. His hand faded into the glowing white several inches before he could feel solid contact. Looking down at his feet, he noticed they too were submerged a little in the white haze.

Backing away from the wall, he stood in the middle of the room, turning slowly while he looked around.

Suddenly, a vertical line grew down the wall, which slowly opened to the size of a door. A small figure of a man came through the door, an extremely old man, wearing a long white robe. He walked slowly toward David, and David began stepping backwards as the old man got closer.

"Do not fear me, David Adams, I will not harm you," said the old man, with a soft peace in his voice. Still in shock, all David could do was stare at him. The old man continued, "Your arrival has been long awaited." The smoothness in his voice eased David a little.

With a trembling voice, David asked, "Who are you, and where am I? Is this heaven?"

Smiling, the old man said, "Ah, heaven, that place where the spirit rests after death. No, my friend, I'm afraid you are still very much alive." It took David a moment to absorb what he had just heard.

"Then what's going on? Where am I, and what's happened to my ship?" David shouted in terror.

"Your vessel is safe, as you are. Please come." The old man then turned and slowly walked toward the door.

David could not move; he just stood there with his back to the wall. The old man turned back to him and said, "Please, do not fear, my friend. You are quite safe, I assure you. Please come." He turned again and walked through the door. With as much self-control as he could bring forth, David slowly pushed himself from the wall and followed him.

Walking out of the room, David found himself inside a long oval corridor, which stretched as far as he could see. Once again he began to tremble. That glow of white was everywhere. He quickened his pace to catch up with the old man, who was walking slowly down what seemed to be an endless hallway.

Stepping in front of the old man's path to face him, David asked, "Would you please tell me what's happening here, and who you are?"

The softly speaking old man simply said, "You must calm yourself. You will have your answers in good time," and continued his pace as if no one was standing in his way. David stepped aside to let him pass, still in disbelief.

They walked together down the seemingly endless corridor. All David could see around him was the white haze. Occasionally he would glance down at the old man, but the old man would just stare straight ahead, with a look of peace surrounding him. It seemed like an eternity passed before they came to the end of the corridor.

Another crease formed in the wall, which opened into a doorway as they approached. As the opening increased, a myriad of colors poured through. David stopped, while the old man just kept walking as if nothing had happened. Just before he went through the doorway, the old man turned without stopping and said, "Your caution is unnecessary. You must trust me, David Adams. Come, my friend," and he went through the door. David hesitated for a moment, and then walked in behind him.

They were in some kind of glass domed room, looking ahead at thousands of stars. It was a small room, circular shaped with the chrome colored walls that came up to about David's waist level, before it met the glass which encompassed the rest of the room. The glowing colors seemed to emanate from the outer edges where the wall met the glass. In the center was a small stairway, leading up to a well cushioned chair that looked something like a throne. David was mesmerized by the domed room's magnificent design.

"Look behind you, David," the old man said.

What David saw when he turned, took his eyes a few moments to comprehend and adjust to what they were looking at. Two incredibly enormous cylindrical shaped objects, side-by-side, extending outward away from them for miles. They joined together to form a single body a few miles ahead, then narrowed as the formation curved upward, eventually coming to a summit point, to the domed room they were in.

Expanding into the far distance, David could hardly see the ends of the two tremendously long cylinders. Straining his eyes, he could barely make out they both ended with what appeared to be two massive, striking streamlined fins, towering majestically upright next to each other.

The entire structure they were on was all shining silver, with an almost chrome stainless steel glare. The dancing colorful haze hovered everywhere, just above its gleaming surface.

David muttered, "What in God's name is this?"

He turned back to the old man, who was looking out ahead at the stars, his hands folded with that same peaceful stare, and said, "A vessel of peace."

David looked to the rear again at the huge structure, then back to the old man. "You mean this is actually a ship? A spaceship?"

"Yes, my friend. Welcome aboard the *Merkava*."

David whispered to himself, "This is a dream, and cannot really be happening."

"There are no untruths here. All that you see, I assure you, is reality," the old man replied.

David slowly opened his eyes, as the old man continued, "I am Sarahan. You must forgive me for interrupting your journey, but when I determined that your course would take you out of the solar system, it was imperative to stop you."

"Are you an alien? Is this an alien ship?" David asked.

"Like you, I am of the Earth. But this vessel is from a world far distant from here," Sarahan replied.

"So life does exist on other planets?" David asked.

"Yes, my friend. Life of which you could not imagine. Beings of such beauty. Civilizations span throughout this galaxy. Worlds with life that have existed long before … "

Sarahan paused and looked toward the window. With a slow pace, Sarahan walked to the stairway and up to big chair and sat. Something was happening. David followed him to where the steps began, next to the stairway guardrail.

The haze of colors surrounding them began to somehow come alive. David started hearing a low-pitched moaning roar, as the field of colors began to dance in and around the great ship, especially toward the far end. As the low thundering roar increased, the swirling colors glowed brighter and moved faster, interchanging through each other faster as each moment passed.

Something else was happening at the same time, but it took David a few moments before he realized what it was. The field of stars was moving.

Turning back to Sarahan, he asked, "What's going on?"

Sarahan seemed to be in some kind of trance, his eyes barely open. Speaking so softly in almost a whisper, he said "Unfortunately, there are beings of not so much beauty; such as ones of greed. You must excuse me for few moments." Sarahan closed his eyes.

CHAPTER 4

INTRUDER

"Sarahan?" David asked, but the old man remained silent. "What beings?" Still no response.

David backed away and looked outward to the domed glass. The stars were still moving slowly, some a tiny bit faster than others. "We're moving," he thought.

He looked to the far end of the vessel. The sparkling colorful blaze above the ship's surface gleamed even brighter now than he had seen yet, with streamlined patterns that flowed alongside the ship toward the aft end, dissipating just beyond the towering fins, then re-emerging into a wake trail of beautiful blue haze that seemed to stretch on forever.

"Magnificent, isn't it?" David heard from behind. He turned and noticed Sarahan had come out of his trance, staring ahead at the stars.

"Are you all right?" David asked.

"Yes. I was attempting communication," Sarahan replied.

"Communication? With who?" David asked.

"A potential intruder," answered Sarahan, still staring ahead.

"Intruder? What do you mean, intruder? Where?" David could not get the questions out fast enough.

"There," Sarahan said, as he pointed outward to the stars.

David quickly turned and looked. In the center of the direction they appeared to be heading for, was a pinpoint of bright orange light. It began to glow brighter and started drifting off to the left, then faded away into the stars.

The roar from behind increased even more, and the streamlined colors glowed now with such brilliance, David had to squint his eyes.

The stars began sweeping from left to right, and the little orange light reappeared, now at the center of their view. A small white disk emerged just to the right of the orange light. Within seconds the disk tripled in size, and David began to notice something around the disk that seemed very familiar.

The disk was growing so fast, he didn't have time to realize what he was looking at. David slowly began walking backwards as the giant planet encompassed the whole viewing screen.

The orange glow remained in the center of view, but David forgot about what Sarahan kept referring to as an "intruder", as he was overcome by the rapture of Saturn's enormous size.

The intruder suddenly dropped from sight again, down and to the left. David could hear the thunderous roar from behind him again. In an instant the giant ringed planet shifted to the upper right of the dome's view, now with the rings directly in their path, and the trespasser headed straight for them.

The tiny orange light passed through one of the lines that separated the rings, and seconds later David watched the rings go from under then above them, as they passed through the ring's great majestic plane.

Shifting again, this time down and to the right, the intruder swung around and under Saturn, disappearing behind it.

With another brilliant flash and dynamic roar from behind, the *Merkava* stayed in pursuit.

Emerging on the other side, swinging up from beneath Saturn, they passed under the gigantic belt of rings, headed out to space again. The intruder once again appeared in their center of view, getting closer to them, indicating they were quickly overtaking it.

Suddenly a strange glistening spark of light jumped out from the intruder and hit them within seconds. With a flash of blinding magnitude, David grabbed the guardrail of the steps, turning his head downward, closing his eyes.

"What's happening?" David quickly snapped.

"We are being fired upon," Sarahan said.

Another explosion followed, with the same blinding brilliance, then a third. What surprised David was that the brilliance of each blast did not seem to affect them in the least. Not even a minor tremble.

Apparently Sarahan had enough. The hazy sparkling colors that engulfed the ship began to dance faster, especially toward the rear, followed by a deafening roar from the engines. Moments later they were right in the intruder's wake, almost on top of them.

Within seconds, what seemed to be a more concentrated band of colors burst out from both ends of the enormous ship, jumped immediately toward each other to collide at the ship's center, and shot an unbelievable bolt of energy in the direction of the intruder.

All of a sudden the whole universe lit up like a supernova going off right in front of them.

David quickly turned his head again, with both hands still firmly gripped to the step's guardrail.

After a few moments, the blazing inferno dissipated into the emptiness of space. David's eyes slowly opened.

"It is over," Sarahan said, with sadness in his voice. David staggered to the forward part of the viewing dome, staring outward into space in total astonishment.

"Incredible; absolutely incredible," he whispered to Sarahan.

He turned back to Sarahan, who was still perched in his chair, looking down into his lap, as though exhausted.

"They were destroyed?" David asked. Sarahan casually nodded yes, still sitting down. David looked out toward the stars again, then walked back to the guardrail, and looked up to Sarahan.

"What's wrong? Are you all right?" David asked.

Sarahan slowly climbed out of the chair, and walked down the stairs and to the side of the dome, looking outward. He spoke with that peaceful tone again, "The taking of any life is not a pleasant thing. Come, David." He turned and walked to the wall at the rear of the dome. A crease formed into a doorway as he got closer, and walked through with David behind him.

CHAPTER 5

ENLIGHTEN

David quickened his pace to catch up to Sarahan and asked, "Would you mind telling me what's going on here, and what that was all about?"

Sarahan spoke as he continued to walk, "Our solar system is forbidden to outsiders. They would not heed my warnings, nor acknowledge me in any form, which can only mean they were headed for Earth. It was imperative to stop them."

"You mean destroy them. Why? Was blasting them to pieces your only choice? From what I saw, they were no match against this ship, or whatever it is."

Sarahan stopped and turned to David. "To ensure their return back to the outside cannot be accepted."

"By killing them?" David asked. "When they saw you, they were trying to escape, and after seeing this monster, they were probably so scared that they would never come back here. Why did you have to kill them?"

"You will know in due time, my friend, why no outside trespasser with the intention of getting to Earth, must never be allowed to escape. The reason would too much for you to comprehend for now. Your patience will earn you that knowledge."

"It wouldn't be too much for me to comprehend why they have to die," David said.

"Please do not judge this as an unmerciful act, David. If it will help your apparent resentment for this action, then for the present there is something I can tell you. It is already known throughout this universe, to all civilizations, before they even commence this undertaking, that any attempt to approach Earth will result in their destruction. Yet there are those who will still try, already knowing what fate will befall them. Come, my friend."

Sarahan turned and resumed his walk. David just stood frozen, with that look of shock for a moment, then followed him.

They continued to walk down the seemingly endless corridor. David was trying to think which one of the thousand questions he wanted to ask next.

"Can you at least tell me where this ship comes from?"

Sarahan replied slowly, "From a race of beings far superior to ours. A civilization so advanced that we could not even begin to grasp their intelligence."

A few steps ahead, another crease formed on the right side of the corridor wall that opened again into a doorway.

"You must be hungry," Sarahan said, turning into the doorway. With the events that David had been through since he met up with Sarahan, food was the last thing on his mind. Feeling empty and mentally drained from the past shocking experiences, he felt that maybe something to eat might help him recover back to sanity a little.

He followed Sarahan into the room, similar to the one he recovered in when he first awoke. This one was a little different. It had a small table in the center, connected to the floor by a single stanchion. In front there was a long, rectangular-shaped viewing window, with a spectacular view of Saturn, slowly drifting away from them. Sarahan sat at the table, then gestured for David to sit on the other side. David slowly sat down, not taking his eyes off the breathtaking view before him.

His attention was slowly caught by what was happening on the table surface. Those same sparkling colors began to dance over the table's top for a few seconds. When they faded, David saw a meal fit for a king. He began to feel his hidden hunger emerge. After taking a several bites, he realized this food had a flavor greater than anything he had ever tasted.

More curious about what was happening to him than being hungry, he tried to ask more between bites. "Can you tell me about these beings?" David asked.

"Beings of peace, whose technological advances far exceed any civilizations yet known," Sarahan said, then casually continued with his meal.

"What do they look like?" David asked.

"Similar to ourselves, actually. The same height, two arms, legs, eyes, ears; however a smaller nose and mouth. Their internal organs are very much like ours, only minor differences. Even their heads are shaped like ours, but with no hair anywhere on their bodies. Although their skin has a gleaming bluish tint to it, which is rather beautiful, once you get used to looking upon them.

"But one unique thing I can tell you about them is how they communicate. They do not speak as we do. They reach directly into your mind, with what I would presume to be some sort of telepathy, except much more. You do not hear words, but you can actually feel what they are conveying to you. And when you are in contact with them, there is a joy and peace that overcomes you like nothing I could ever describe."

David had stopped eating halfway through Sarahan's description, trying to fantasize in his mind what they were like, then said, "Are any of them here? On this ship?"

"You and I are the only beings on this entire vessel," Sarahan replied. David just stared at him in awe.

"Then how…why…?"

David could feel himself slowly slipping into that blank state of astonishment again, but before he could stutter again, the old man said, "All the details you will know in good time, David. This is why I can only be vague for now. However, I can point out something that will interest you for the moment."

Sarahan rose from the table, picking up his drink at the same time, and walked toward the viewing window. Looking out at the universe, he continued, "Have you ever wondered why reports on Earth of unidentified flying objects, or UFO's as they are labeled to the general population, have never actually manifested their existence, much less no form of physical proof of them has ever been produced? The creators of this grand vessel prohibit any outside contact with Earth, of any nature."

David thought for a moment, and then said, "Yes, what you say is true, no concrete proof has ever been produced to actually expose their genuine existence."

Sarahan went on, "All the reports ever recorded, were only from those who, in fact, wanted or wished to believe that they had encountered some type of alien spacecraft. Some believe they were truly abducted, to be hosts of physical experiments. All conjecture,

21

my friend. I can assure you with all sincerity, no outsider has ever come within sufficient proximity of Earth to be observed, much less ever close enough for any kind of incarceration, with the exception of the creators of this vessel. They have incorporated a shielding that ensures that no instrument on Earth can ever detect any of their vessels. The civilization that created this magnificent vessel, discovered this world thousands of years ago, and has safeguarded its natural evolution from outside interference ever since."

David got up from the table, walked over to Sarahan's side, and said, "But why? Why does this vessel protect the Earth? With a race that is far superior to ours, what's so important about protecting that God-forsaken place?" David began to feel all the hatred he left behind, slowly begin to resurface within him.

Sarahan looked at David with curiosity, and said, "I perceive anger in your words. Why do you speak of Earth that way?"

"Why? I'll tell you why!" David stepped closer, and was face to face with Sarahan. "Because every form of selfishness, hatred, and every other evil you can think of, exists there!"

Pointing out of the window, his voice began rising as he continued, "People are so wrapped up in their own pathetic lives, all they can think of is who they're going to stab in the back next, just to gain something! I've dedicated my whole life to developing a new type of energy that would revolutionize space flight, while my so-called 'colleagues' laughed at me behind my back! And where are they now? Now that I have made it come true, I have finally rid myself of that race of stagnating parasites, and what happens?

"You come along with this…this 'ship', snatch me away from my journey away from that ball of malevolence, and then tell me that you're protecting it? Well, if it's all same to you, I have a galaxy to explore, and I would like to be on my way now, thank you very much!"

Sarahan turned and walked back to the table. He sighed and turned back toward David, and said, "You cannot."

David took a couple of steps toward him, and shouted once again, "What do you mean, 'cannot'? What am I supposed to be, your prisoner now?"

Sarahan slowly eased down onto the seat by the table, looked up at David with a surprising half-smile. "No, my friend, I do not hold you as captive. To continue on your journey would be extremely dangerous once you got outside the solar system. There is a good possibility that you would encounter beings like the ones we just dealt with, and your journey would end very swiftly."

David kept having this obsession during all the conversations, to keep gazing out of the viewing windows, as if he were still trying to accept the fact that he was really here. He did so again after he finished listening to what Sarahan just explained.

After a few seconds, David spun back to Sarahan, and with a touch of more fury, shouted, "I'll take my chances, thank you again."

The old man lowered his head as he spoke, "Do you not understand what I am trying to say? What you have labored and dreamed for all your life would be wasted. Your burning desire that I feel within you will all be in vain. You have come this far in your quest; do not jeopardize your life by exposing yourself to the malevolent inhabitants of this universe."

David thought for a moment, and then said, "It's not the evil unknowns I want to see, it's the beautiful ones. Besides, I didn't come this far just to stop here, much less go back." David sat down to face Sarahan, and continued with a softer tone. "Look, I appreciate your concern, and for whatever reason you're protecting that God forsaken planet, which I don't really care about now. All I know is I just want to put as much distance between me and them as possible."

Sarahan looked up at David with despair, and said, "One so intelligent, yet driven by anger. Very well. I cannot hold you against your will. Come then, I will not detain you further." Sarahan slowly rose, turned and walked toward the wall, which formed into the same corridor, and walked out. David took a last glance around the room and out at the stars, and followed Sarahan down the corridor.

The two did not speak as they walked the hall chamber. David thought they would be walking again for miles, but after a short distance Sarahan stopped, but did not move.

After a moment, the floor in front of them began to part, just as the doorways did when they formed in the walls.

Sarahan started walking again toward the middle of the parting, as if he was going to fall into the opening. As he passed the edge, he began to descend. David noticed he was beginning to walk down some kind of stairs, so David shrugged a moment, and followed him.

Sarahan was moving downward toward some kind of wall at the bottom, which David knew was going to crease into another doorway when the old man got close enough. It did, which brought them into a much larger room than any he had seen since he had been aboard.

This room was as big as an auditorium, with a ceiling that seemed to reach to infinity. While taking into account the size of the enormous ceiling, David heard Sarahan say in a low voice, "Behold."

About 10 meters in front of them, stood David's little ship, gleaming like it was surrounded by floodlights, even though there was no source to the lights.

"My control will take you out and safely away from this vessel, then you may engage your engines," Sarahan directed him. David just stood there, not knowing what to say. Sarahan turned to face him, then continued, "I understand your feelings and your desire to flee from them, but I do not believe all men are as you say."

David lowered his eyes to the old man and said, "Understand? How can you understand? How can anyone ever know what I have been through to achieve what I have? You can 'understand' what I feel, but you will really never know."

As if Sarahan did not hear a word that David said, he slowly smiled and said, "May your journey be without peril, and hopefully one day you will find peace in your heart to forgive your fellow man. Goodbye, my friend."

Sarahan slowly made his way around him, and walked through the passageway that sealed up behind him. Staring at the now-blank wall for a moment, he walked to his little craft.

Once he opened the hatchway, David took one final look around the tremendous room. He felt a sense of guilt for leaving Sarahan, that he was deserting him. David sealed the hatch and walked through his ship, checking all the instruments. Everything was in perfect order, as he knew it would be, for some reason.

Opening the view port shutter, he noticed the colors began to engulf his ship.

He climbed into his pilot's chair and strapped himself in. The colors started sparkling brighter, and he began to move. David watched the wall through which they had come into the huge room drift away.

The little ship turned, and the front of it was now in the direction of motion. Further ahead, David could see a giant crease forming in the floor, then part away from each other to form an opening. As the opening expanded, he could see the background of stars.

His small ship sailed through the opening, out into space. David unbuckled his safety straps, jumped out of chair and quickly walked to the rear viewing port. The great size of this alien vessel overcame him again. David watched, as it slowly drifted away, until the great ship was finally out of view.

The surrounding colors on his hull vanished. Now he was on his own again. He turned on the main engines, and engaged the navigation and thruster controls.

There was a comfort in hearing the low hum of his engines, knowing he was on his way again into the security and solitude of space. He eased back in his chair, thinking of all that just happened while on that incredible ship, wondering now if it was all some kind of a dream.

Though his hatred would not cease for the rest of humanity, he could not help but feel a certain respect for what Sarahan was doing; what still confused him was why. He wished he could have spent more time with him, but his desire to explore the stars was too overwhelming.

David knew he could have probably learned everything about the mysteries of space from the old man, but discovering those unknowns for himself was a big part of his lifelong dream.

One of the greatest unknowns that mankind speculated about for centuries, had now been revealed to him, the fact that other life did exist in the universe.

Yet again, there were unanswered questions that kept creeping up in the back of David's mind.

Such as this alien civilization, with the intelligence and technology so far superior to the other inhabited worlds of the cosmos, especially the human race. What was so important to them about protecting Earth?

Not to mention these "intruders" that were virtually attempting nothing less than suicide missions, just to get to Earth? Surely there was a great deal more there they were concerned about, than their own curiosity.

But there was one question that gave David the biggest sense of mystery above the rest.

Why were these alien architects, of such an advanced race so far superior to humanity, using a human being to pilot and direct their inconceivable vessel, and not themselves?

David kept trying to put these puzzles aside, by reassuring himself that anything to do with the rest of the human race, was no longer his problem, and continued on his voyage.

CHAPTER 6

SEIZURE

Checking his instruments, he knew he was beyond the orbit of Pluto, well out of the solar system. David locked his navigational instruments for Alpha Centauri, the closest star to the solar system, thinking it was as good a place to start as any.

He could now engage the velocity controls to full power, which would increase his speed far past the speed of light. Grasping the throttle control, he hesitated for a moment.

David looked back towards the sun, now the size of an ordinary star, lost in the vast ocean of the cosmos.

He envisioned this moment so many times on Earth, setting outside the solar system, looking back at the sun, wondering about the adventure ahead. Little did he know that his adventure would start just after he left Earth. But like his life on Earth, and his encounter with Sarahan, it was all behind him now.

Turning his attention back to the instruments, he slowly moved the velocity control up to full power. The pulsing roar of the engines began to make his ship tremble, and his speed increased.

According to his calculations, it would take about three weeks to reach Alpha Centauri. Three weeks, he thought. He eased back in his pilot's chair, still trying not to think of all that had happened the last few days. Exhaustion finally took its toll over him, and he drifted off to sleep.

Dreams slowly began to fade in again. Beauty he would encounter, beings he would meet. Sometimes Sarahan's tremendous vessel would focus in and out of his dreams, cutting across his path, warning him to not go any further.

He woke up sluggish. David glanced at the clock on the control panel, which told him he had been sleeping for about 4 hours. He climbed off the chair, and walked up to gaze out of the forward viewing screen. Again he was lost in the rapture of the never-ending universe. This was one sight he knew that it would be impossible to get tired of looking at.

It was boredom that David thought was going to be a problem, but a view like this, and the anticipation of an adventurous journey ahead, he knew would slake a good part of the boredom.

Suddenly, the sensor alarm split through the silence. David jolted up, went over to the control panel, and switched off the alarm. Tuning in the sensor screen, he saw there was an object just ahead. It was another ship, except this was only a little bigger than his. Quickly looking out of the viewing screen, he could barely make out a blue hazy shape, slowly getting larger. Glancing back at the sensor screen, David began to get concerned. The object was closing on him fast.

He jumped to his chair, grabbing the controls. He turned hard right, and could feel his little ship begin to shake. Within seconds, the whole interior of his ship was flooded with a blue-white haze, which instantly paralyzed him.

He tried to move, but could not budge a muscle. All he could feel was his heart beating so hard in terror that David thought it was going to jump out of his chest.

Quickly growing thicker, he felt like something was smothering him. He strained to breathe, but couldn't move his lungs to inhale. Vertigo rapidly overtook him into unconsciousness.

David felt as if he were drifting in some kind of cloudy endless void, unbound by any physical restraints. As he floated along, he began to feel something touching him, not from outside, but from within. A spinning motion of some sort was taking place within him, yet he could not see any part of his body.

The spinning sensation started to slow down. He began to hear sounds. They were like voices, but were much more erratic. The sounds echoed from one tone back to another, unlike anything he had ever heard. He tried to locate the source, but could not seem to get a fix on it.

The spinning sensation finally stopped, and felt stable, however he could still sense something moving within him. Warmth started making its way into his awareness. Lights started piercing into his focus, and the erratic sounds were becoming louder.

When his view finally cleared, he realized he was in an upright position. There were two figures standing before him. They looked humanoid, about the same height as David, except much thinner.

Their faces were extremely pale, almost a ghostly white, with what appeared to be small ears and long slits for eyes.

The one on the left of him had his right arm passing what looked like an instrument over his forehead. The being next to him appeared to be just observing. As the instruments passed over every part of his head, David felt pressure at whatever part of his mind the instrument was moving over.

Though some of it was unclear, David began to remember what had happened. His ship had been overtaken by something. Another ship? Perhaps, and now was being held captive. His awareness increased, and fear started to overcome him.

Attempting to lurch forward, pain shot through his back like a burning stake jabbing into his spine. Jerking his head back, he tried to scream, but couldn't make a sound. The being with the probe instrument was startled, and jerked his probing mechanism back, while the being on the right quickly took a couple of steps back, as though both were in fear of him.

Something held him in place. The more he tried to move, the greater the pain increased. He was forced to ease off his effort to move, which caused the pain to subside. Both aliens slowly crept back for a look a him, chattering something. Neither one would get as close to him as they were before, as if they feared him for some reason. They turned and walked off in the same direction, still speaking to each other in that chattering language they spoke.

All David could do was lay there, paralyzed in horror. His heart was pounding so fast he thought it was going to explode. He even tried to shift his eyes for a look around, but even they would not move. All he could do was stare straight ahead.

This can't be happening, he thought. To come all this way, only to end up as a zoo specimen. He was completely helpless. Tears began to roll down his face. The old man was right. He had been warned, but it was now too late. He wanted to seek the unknown, but in turn the unknown sought him into captivity.

CHAPTER 7

LIBERATION

After a few moments, David heard them speaking again, and opened his eyes, only to find the same two beings standing over him again. They had no instruments this time, just pointing at certain points of his head.

All of a sudden a third being quickly stepped into his view, chattering much faster and louder than the other two had been. All three dashed off out of David's view, as if panic stricken.

Seconds later, a massive jolt shook everything around him. Whatever was holding him in place released him, and he collapsed to the floor. Raising his head up, every light in his view went out. He tried to move, but it was as if every muscle in his body weighed a ton.

Throughout the pitch black around him, he could hear those chattering voices now screaming, as if they were terrified. Some kind of pressure started to close in on him. He tried to breathe, but there was no air to inhale. He felt as if he was being crushed to death, as the pressure around him got worse by the second .

The end had come, so he closed his eyes to meet his death. His journey beyond his wildest expectations was nothing but darkness now.

Only a few moments passed, and the pressure began to ease off a little. Air began to fill his lungs. As he opened his eyes, all David could see was that he was in some sort of void, with colorful lights dancing all around-lights that he recognized. They swirled and danced in and out of his sight like sparkling gleams of joy.

The crushing pressure was all but gone, however every muscle in his body felt as if it had been beaten and bruised. He was breathing with relief now, but not deeply. The vertigo feeling subsided, and the dancing lights began to fade away. As the last of the lights disappeared, David could see a small figure standing over him, with that same peaceful smile of Sarahan.

"Hello, my friend," the old man said. David reeled his vision from side to side, realizing he was back in the same room as when he first came aboard the *Merkava*, all glossy white with that misty haze hovering just above all the surfaces. He tried to sit up, but cramps throughout his body refused to let him move.

"Remain still, David. You will be fine in a few moments." Sarahan lowered his head and closed his eyes. Out of nowhere, those amazing colors seemed to emerge and encircle David's body. They danced and sparkled about him in an array so brilliant, it was almost blinding.

David could feel the pressure of the cramps slowly dissolving, and that same warming peacefulness began to flow through him. He turned to the side Sarahan was standing. The old man was still in his trance.

"What happened? Who were those, those beings?" David muttered.

Without opening his eyes or raising his head, Sarahan whispered, "You were abducted." David rolled his head back, looking straight up again.

"By who, or should I ask, what?" David could slowly feel his strength returning.

"Please remain silent and still for a few moments," Sarahan whispered back.

The colors that swarmed around David were now beginning to disperse and fade away. As the last of the colors vanished, Sarahan looked up at David and said, "You may sit up now, but move with care."

David carefully raised himself up and let his legs drop over the side of the table. He could still feel his joints aching with a little stiffness. He tilted his head back, taking in a deep breath, and slowly released it as his head came down, looking straight into the smiling face of the old man.

"What did they do to me? I was completely paralyzed, because it felt like I was being crushed to death."

Sarahan replied, "You were restrained by a heavier air pressure method. I have compensated, but I'm afraid you will be a little uncomfortable for a short time. In simple terms, you will undergo a slight case of what is now called on Earth, the 'bends'. You will be all right shortly."

Then he handed David a small flask, containing some kind of silky fluid. "Drink this, it will give you strength." David looked down into the liquid, and noticed tiny glitters sparkling on and off.

Hesitating, he looked back up at Sarahan, who smiled at David and said, "It is quite safe, I assure you." David took a small sip, and immediately started feeling a sort of warm energy surge down and through him.

"Not bad. What is it?" David inquired.

"The physical properties would be difficult to describe," Sarahan explained. "Trust only for now that it fully restores our health."

"Well, bottoms up," David said, and drank the rest.

It was if new life itself was springing throughout the rest of him. He let himself slide off the table, so that he could attempt to stand, and found no trouble in doing so. Sarahan, already moving toward the newly formed opening in the wall, turned and motioned David to follow.

They made their way to what David now thought of as the "bridge"; the domed room at the extreme front tip of the *Merkava,* with only a large cushioned chair at the top of a small set of stairs raised in the center. Sarahan began to climb the few steps leading up to it, but stumbled for a moment and grasp the guide-rail to keep from falling.

David quickly jumped to support him, then asked, "Are you all right?"

As Sarahan straightened his stance, with that same smile he said, "Sometimes my age catches up with me. Thank you, David."

David made his way to the front of the domed room, staring outward, thinking of the nightmare he had just been through. He turned back to Sarahan. "I thank you," David sighed, "I never want to go through that again. Just who were they?"

Sarahan looked back to the stars. "One of many. I knew your fate before you departed."

David turned back to him, and asked, "Then why did you let me leave?"

Sarahan turned his eyes away from the stars to David. "Some men do not learn from the word of other men. With your anger that blinds you, had I not let you go, you would always doubt me."

"True." David said.

"Also," Sarahan continued, "If you had ventured out much further, I would have been forced to interrupt your journey, and retrieve you, as I knew this would happen. It is imperative that no outsider captures a human being. There is a point outside this system that I cannot travel beyond, for fear of not being able to return to Earth in time to intercept any potential intruder. I was hoping this would happen before you got much further. Fear not, my friend, for I knew I could get to you safely before any ship had time to escape with you."

"I am really grateful you kept tabs on me. I'm certain that was the only way I would have learned, too. But I've dreamed about exploring the stars all my life, and worked so hard to make it happen. It's just not fair that I've come all this way, only to find out no one can leave our solar system, because of a bunch of headhunters.

"But now my curiosity has changed. I don't understand, and would now really like to know, what is so important about protecting the Earth? What could any race, capable of space flight, meaning they have to be more advanced than the human race, possibly want with Earth?"

Sarahan rose from the chair, walked on the steps to face David, and said, "My instructions are to inform Earth, once they have achieved interstellar space capability, of the danger. Also to let them know why. Since you are the first, even though you are years ahead of your fellow man with this technology, and do not wish to share it with them, does not alter my guidelines."

Sarahan looked again toward the stars, reluctant to tell David. But David was persistent.

"Well, let's have it. What's so important about Earth that these outsiders, knowing they are going to get blown out of the sky if they get so much as even close to it, still try anyway?"

The old man turned back to David. "It is not the Earth that interests them; it is us they seek."

"Us?" David asked in amazement. "You mean humans? The human race?"

Sarahan slowly motioned yes.

Walking over to his side, still confused, David exclaimed, "You've got to be kidding! Now I'm even more confused. That makes no sense whatsoever. What could they all possibly want with the human race?"

The old man rolled his sagging eyes up to David, and replied, "Much, my friend; very, very much."

CHAPTER 8

JUSTIFICATION

David thought for a moment, letting what he had just heard sink in, and then replied, "For instance?"

Sarahan started to speak, but hesitated for a moment. His gloomy, sagging eyes then came to life. He walked up to the big chair and sat, still silent. David had seen that look before.

"Another one?" David asked, and then looked out to the stars. Must be, he thought, since Sarahan was in that semi-sleep trance again.

"Yes," Sarahan said as he lifted his eyes, now fully alert. "Two vessels are approaching the breaching point."

David could feel the gigantic vessel surge with power again, while the star field view outside the dome began to leisurely shift. Looking to the rear, David saw what he expected to see, those two cylindrical ends of the ship had already burst into those brilliant dancing colors, and the ship swiftly began to gather speed.

"Where are they now?" David insisted, with a sharp tone of annoyance.

"On the other side of the solar system, heading inward. We must make haste."

The enormous *Merkava* was doing exactly that. David watched in amazement. Saturn swept past their right side in the wink of an eye. Looking ahead, he saw the small glowing disc of the sun, rapidly growing bigger. Instinctively he backed up, in total rapture of the tremendous fireball rapidly filling his vision, knowing they were about to plunge directly into its heart.

David screamed, "Have you gone mad? Turn this thing!" Backing up so fast in a state of panic, he felt the rear wall of the domed room slam against the back of his head, and yelled again, "What are you doing?" David shut his eyes, as the sun's blinding brilliance now filled the entire view of the dome.

David's back slid down the wall, almost to the floor, and shouted, "You're going to kill us, you crazy old man!"

But to his surprise, he heard Sarahan speak in that same level-headed calmness, "You should calm yourself, David. To worry is not to gain, my friend."

Still in his catatonic state, he yelled back, "This is not the time for useless phrases, you old nut! Not when we are about to be barbequed!"

"I had hoped you had learned to trust me a little more. Please look," Sarahan said, with his tranquil composure.

David opened one eye, and saw the curvature of the sun's surface promptly moving under them. Just as he got to his feet, still trembling, a solar prominence jumped into their path, a tremendous display of a super pyrotechnic explosion, swelling up right in their path that would have engulfed a thousand Earths.

Frozen with shock, all David could do was watch. The magnificent *Merkava* spearheaded the prominence with incredible velocity, bursting directly through it, like a bullet blasting through an ocean of hot lava, sending plumes of fire in every direction for thousands of miles, pulling some through their wake of invincible momentum. David quickly turned to the rear, and watched the immense statue of fire settle back into the sun in a magnificent splash.

"Good God Almighty!" David mumbled, watching the sun shrink back into a small glowing disc, while they streaked further on.

The field of stars shifted to the right, as Sarahan banked the great ship to the left, in full pursuit of their new intruder. David looked up to the chair, only to notice Sarahan was back in his trance. The unbelievable speed of this vessel was something David still could not fully comprehend.

The ship began to slow down. David observed two small pinpoints of red light ahead, hovering side by side. "Is that who you're after, or should I ask, they?"

Sarahan still remained in his trance. David looked back at the two objects, and noticed they were fading away. Looking back up to the old man, Sarahan was fully awake now.

"Who were they?" David asked.

"Peaceful travelers," Sarahan answered with a sigh of relief, while walking down from the elevated chair. "Just explorers, a little off course. I know their race, one of whom I would one day hope you will give me the honor of introducing.

"However, all vessels that come near to our system, must be investigated. You see, there is one race that could present a severe danger. A powerful civilization known as the Caperians. They have acquired the technology of stealth, even to the magnificent beings that created this vessel. The only way I can detect their presence, is when they come within only a few thousand kilometers of Earth."

"Then how do you detect their presence, if they have a method that makes you unaware of them?"

"Let's just say for the moment, when their approach brings them within close proximity of Earth, I have the ability of 'feeling' their presence."

David just stared at him with confusion.

"Come, my friend, and let us refresh ourselves." Sarahan walked out of the domed room.

David, still in somewhat of a cold sweat, was looking back at the sun. He said as he was turning to follow him, "You wouldn't happen to have a bottle of Scotch on board? Sarahan? Wait up!"

CHAPTER 9

DISCLOSURE

The ship was moving again, but toward the outer region of the solar system. Soon they approached a cluster of asteroids, but the *Merkava* did not yield their presence by slowing down. Instead, just when the *Merkava* closed in on them, the huge tumbling rocks parted a couple of miles ahead of where the front of the ship came to a point, as if some sort of invisible shield ahead was creating a pathway.

After a few minutes, the ship slowed to a stop, and turned completely around to face the solar system. Millions of these asteroids surrounded the great vessel, rolling and tumbling about, but did not come near the ship.

David and Sarahan ate in what David remembered as the room he had a meal in when he first came aboard. He watched the rolling space rocks calmly turn and softly move in the void of space.

"The Ort Cloud, you say. Quite a place. So this is what's left from the origin of the solar system, and where our comets come from. Like that nasty one that wiped out the dinosaurs. Do you stop them, too?" David asked.

"Only if they are a threat to Earth," Sarahan answered.

"I should have known. Why did I even ask?" David said with a touch of sarcasm. He went on, "I know that I'm the one who has to bring this up again, because I know you're not going to voluntarily tell me."

"Yes, my friend. As to why the outside civilizations wish to obtain the human race. Or even one human being. I have been contemplating your answer, and do not know if you are ready for the answer yet. It is not beyond your comprehension, but I would believe it is well beyond your understanding," Sarahan said.

"Would you stop with the riddles, and just tell me?" David snapped back.

Sarahan rose from the table and walked to the window, looking out at the nomadic asteroids. He turned back to David and slowly replied, "All beings of intelligence in the known universe are born and live their entire lives, using every available portion of their minds."

He slowly walked and looked up to David, and said, "With the exception of one."

Turning away to the stars from David's confused expression, he continued, "Humanity, David. Only the human race employs between ten to fifteen percent of their minds. The other eighty five to ninety percent remains inactive, undeveloped."

Now more confused, David sat down. After staring blindly for a few moments, he looked to Sarahan. "I've heard this, but no one has ever come up with a clear-cut answer as to why. I don't suppose you know?"

Sarahan sat down with a heave and a sigh. "No, I do not. In that sense, I am just as puzzled as you are. As are the creators of this glorious vessel, and every other advanced life form that knows us."

He sat up and walked to the very point of the domed room, looked out among the heavens, and continued, "But there is something they, I, and every intelligent race in the known universe that has achieved space travel, do know. The human race has the most rapidly evolving mind in existence."

He turned to David. "The architects of this ship took millions of centuries to develop from their discovery of fire to their present existence. The other civilizations that have the capability of space flight took even longer."

He started walking in David's direction, and went on, "It has only taken a little less than two million years for our race to evolve from the development of fire to our present awareness. The human race harnessed electrical power only a little more than a century ago, along with the very beginning steps of grasping nuclear power. Nuclear fission, the splitting of atomic particles, came first.

"Now that you have discovered the next step, the human race is now on the verge of learning to control nuclear fusion; the joining of atoms. It took a millennium for other cultures to grasp the knowledge of atomic physics.

"Billions of other intelligent life forms, much more advanced than ours, perished throughout the universe attempting to grasp the might of nuclear energy. Once the way to harness it was finally discovered, the knowledge itself was the greatest treasure in the universe, and billions more died in warfare amongst some civilizations, just for the knowledge of how to split atoms." Sarahan practically had to catch his breath, revealing this to David, who sat listening to the old man with astonishment.

Sarahan continued after a brief moment, "Fortunately, Earth was still not discovered during that significant period of time."

Through his look of overwhelming amazement, David asked, "So what you are telling me is that the human race could grow into the most powerful race in the history of creation?"

"No, my friend." Sarahan leaned toward David to look him straight in the eye. "We already are."

David reclined in his chair in disbelief. "What? What are you talking about?" he sputtered out. "I'm the first human, present company excluded, to get this far out on my own. The first. Yet there are God knows how many other races bouncing around the universe already. Speaking of which, how on earth, no pun intended, did you win the prize of controlling this oversized monster? Shouldn't your 'creators' of this ship be running the show here?"

Just as David finished his question, Sarahan rose from his seat with his eyes closing to a slight vision. "Don't tell me. I know that squint. Something's coming, right?"

Sarahan was already heading for the opening crease that led to the corridor, and spoke softly to David, "Come, my friend."

"You're just trying to avoid my question. You can run, but you can't hide." Sarahan was already in the passageway, as David quickened his pace to catch up with him.

CHAPTER 10

CHALLENGE

"Please be silent, David," Sarahan told him as they entered the domed room, which David now thought of as the bridge.

Still, as Sarahan climbed up to his control chair, David persisted. "Not until you tell me what makes you Pope of this palace."

Ignoring David completely, Sarahan sat in his chair and lowered his head into that same trance. The slow tumbling asteroids suddenly began to part, and the ship was underway. Always fascinated by the burst of the enormous stream of blue haze that blazed out for miles behind them at the ship's stern whenever Sarahan engaged the engines, David turned back to look.

When he turned back around, they were already clear of the Ort Cloud, as the majestic view of the colorful star field exploded into view.

Now in open space, they banked hard left, and the ship's power thundered by increasing in velocity more now than David had ever seen, feeling the ship's engine's deeper growl. Watching to the rear, he saw the magnificent blaze of now a solid blinding white inferno shadowed by those soaring huge tail fins. It was followed by the rich blue trail that flared outward perhaps a mile from the stern, like a high altitude jet's contrail, he thought. But those contrails did not begin to compare with the astounding sight David was witnessing now. He had never seen the ship accelerate this fast before, from a virtual stand still, to far past the speed of light, within just a few seconds. This actually gave him goose bumps. David was beginning to really like this incredible colossal-size craft, not that he would ever admit it to Sarahan.

David saw that Sarahan was now out of his trance, but for the first time he noticed an expression of worry on his face. "What's up now?" David asked.

"This is a little confusing. There are three vessels traveling at maximum speed just beyond the orbit of Neptune, but their course is not for Earth, as they are … just as I suspected … decoys!"

Sarahan immediately turned towards the sun. David heard the engines roaring so powerfully now, the ship began to tremble with power.

"My age is starting to show, David. How could I have been so foolish?" Sarahan was actually raising his voice for the first time.

David saw the blue-white disc of the Earth coming up fast, with about ten red pinpoints of light heading for Earth, about halfway between the Earth and the moon. In the wink of an eye, Sarahan positioned the ship directly in their path.

David looked up just in time to watch the colliding colors, already in motion, slam together, resulting in several thin glaring lines jump outward. He turned to the intruders, but the thin rays beat him to it. He watched every red dot vaporize instantly.

"Nice shooting!" David shouted. Sarahan was partially back in his trance, but didn't respond to him, which to David could only mean he still had his hands full.

The *Merkava* was moving at almost blinding speed now, outward away from the sun. He already knew what Sarahan was after.

"Those were actually decoys?"

"Correct, David. They too must be eliminated. This will only take a moment." Sarahan's eyes were fully opened now, focused in a straight line ahead, watching the three red specks of light rapidly come in to sight, then destroyed only moments later.

They both sat in what David noticed a new room now, that had a panoramic view of the stars almost as incredible as the domed room.

"That happened so fast, I couldn't turn my head fast enough. But I think I know what they were trying to do. The first three ships were trying to lure you away from Earth so that little armada could get to Earth. Right?" David asked.

"Exactly, David. This is not the first time a deception like this has been attempted. But it still confuses me that the greedy races still try this maneuver. They all know … every civilization knows … that they are no match for that which protects the Earth. Yet, to this day, they still try.

"But what concerns me even more, my good friend, is that when I first detected them, I clearly knew their course was not for Earth.

"My first trained instinct was to focus on Earth, to see if this was some sort of deception. Yet, there are so many peaceful travelers who accidentally cross the boundary into our system, that I allowed myself to believe they were only that, which distracted me from also sensing anything coming within the proximity of Earth. But with age, comes less concentration, and more wishful hope, such as, will there ever come a day when I do not have to take away other being's life?"

"Just how old are you?" David asked.

Sarahan smiled. "Two thousand, five hundred seventeen years." He kept smiling, waiting for David's reaction, but all Sarahan saw was an opened-mouth expression attached to a frozen face.

After passing his hand across David's catatonic stare, Sarahan caused David to jump up and shout, "Hello! Okay! That does it! Enough is enough! Would you just please start from the beginning, and tell me just what is going on here? I'm so tired of these puzzle parts bouncing around in my head from you dancing around the whole picture!

"I finally leave Earth, all goose-bumpy thinking my dreams are finally going to come true. But no, I get hijacked by you and whoever they were, watch you blow about twenty ships into oblivion trying to get to where I don't want to be, and now you tell me you're older than Jesus Christ! This isn't a dream coming true. It's like I've slaved all my life to have my worst nightmare unfold! Since I passed up crystal ball reading, the universe as of this moment is now officially on pause, before I end up in a straight jacket! So guess what? You're not moving out of that chair until I know exactly what is happening here! Okay?"

Sarahan was grinning even more now, and quietly said, "Sit down, David." He then leaned to the table, picked up a glass which contained some sort of drink, and handed it to David.

With a little hesitation, David took the drink, not taking his enraged eyes off the smiling Sarahan, and carefully sat across from him.

"What is this?' David asked.

"What you asked for. It sounds like you could use one by now."

David sniffed the top of the glass, and to his surprise it was. "Scotch?"

Sarahan nodded yes, then said, "A double, in fact. And I think you will find this brand the best you have ever tasted. Enjoy it, while I give you the … how do you call it … 'big picture', my good friend.

CHAPTER 11

DEPICTION

"Twenty five hundred years ago, I was born and raised in a land which today is called Morocco. I lived alone in a small village surrounded by many miles of desert land. Such a peaceful place it was.

"Every night, I would ride my camel Josh a few miles into the desert, spread a blanket out, and lay on my back looking up at the magnificent heavens, listening to the calm desert winds."

"Wait a second," David interrupted. "Back up a few steps. You had a camel named 'Josh'?"

"His real name was Joshua, which the people in my village thought disgraceful, since I named him after a holy man. It was not my intention to insult anyone. I just liked the name."

"Sounds like a pretty cool name for a camel, not that I thought camels had names."

Sarahan just frowned at David with a cold, blank stare.

David looked to the side, and muttered, "Sorry. Go on."

After rolling his eyes up and down with a heavy sigh, "If I may continue. I was so fascinated by the beauty of the night sky, always wondering what the thousands of twinkling lights actually were. I knew God had put them there for a reason, a reason much greater than just for us to gaze at and admire.

"Lying there, I would raise my hand up, wanting to touch them, wishing so much to go out into this vast ocean of incredible splendor, until slipping off to sleep, only to dream of drifting among the magnificence of the universe.

"One night, I awoke to a gleaming light, assuming dawn was drawing near when first opening my eyes. I then realized this light was sparkling with astounding colors, growing brighter by the moment, surrounding me completely. Before I could even imagine what was happening, my entire vision was engulfed, and was overcome by more fear than you can ever know.

"When my vision cleared, I realized the dancing lights had vanished. But I was no longer in the desert. I was in some sort of room, similar to this one, sitting in some sort of chair. Still frightened, I looked about the room, wondering what fate had come before me.

"Suddenly, a hairline crease began to emerge in the wall directly in front of me. The crease grew into the same opening that you have seen many times here as a doorway. Just when this opening became wide enough and stopped, five beings, of such unique magnificence, leisurely walked into the room. They gazed their glistening eyes upon me as they approached, eyes that looked like those of angels, and they all stopped a few feet before me. I was so terrified, becoming petrified to the point of not being able to move, shivering with horror.

"It was then that the most unbelievable experience of my life took place. In the wink of an eye, David, all my fears amazingly turned to an inconceivable peaceful joy you could never imagine. A feeling of such glorious harmony, I could never describe. You could only know the magnitude of this bliss, by feeling it for yourself."

Sarahan smiled while he rose, and strolled over to the viewing port, looking out to the universe with such ecstasy, more than David had even seen in him.

"Then what?" David asked.

Not turning to look back at him, he went on, "But that was only the beginning, my friend. What transpired next was even more incredible." He turned and walked back to David, and stood before him.

"As this feeling of magnificent peacefulness was still growing inside me, all five beings dropped down to one knee, and bowed their heads before me, in some kind of gesture of worship."

He then sat, facing David's confused look, and continued, "Though this joy was now well abound within me, I could not understand why they were acting as if I were some sort of deity. After a moment, as they slowly raised their heads up to face me, a voice of such graceful beauty that my ears could not hear, but could actually feel, grew instantaneously into my consciousness.

"My awareness could feel, 'We have come from a very distant place, and ask forgiveness for this intrusion. Exploration of other star systems has forever been our only purpose, and we do not disturb the inhabitants that dwell upon the worlds of the systems we observe.

"'When our probes returned after completing the survey of your planet, we encountered a remarkable discovery, and deemed it necessary to extract one of your kind for extensive investigation.

"'We therefore chose to begin with you, and brought you here to continue further examination, which will neither injure nor harm you in anyway.'"

"My original reaction was why me? But when I began to ask this, they had already somehow detected my question. Again, with no words, but what I now know was a highly developed form of telepathic ability, I could feel, "'Our probes further revealed that you, more than any other of your kind, wished to know what secrets lay beyond the surface of your world. By no means will we hold you against your will.

"'If your decision is to return to your world, we will bring you back to the very place we apprehended you, ask again your forgiveness for your detainment, and attempt to select another.'"

"With no hesitation, they felt my answer electing to stay, and how honored I was to be the first of my kind to be presented with their offer of such a magnificent opportunity. They returned my gratefulness with overwhelming gratitude. But most unique part in communicating with them was that every time they responded, it was more than telepathic signals. Oh, I could decipher their reply with complete clarity, but when their words entered my consciousness, my mind felt as if it was developing into a greater awareness, because of their peacefulness, their knowledge, and culture. My mind was actually giving ground to new realities, genuineness, as though it was expanding."

David was becoming more interested now than confused, and asked, "Let me interrupt for a second. Twenty-five hundred years ago, everyone thought the Earth was flat, and everything, including the sun, revolved around the Earth. When did they tell you different?"

"I'm coming to that, my friend. When they first entered the room and kneeled before me, I instantly felt the realism that the night sky I saw for many years in the desert, was not a black blanket dotted with lights, but a massive void with other suns so distant, they appeared as sparkling specks of light. But they were not sending me any messages yet.

"And the closer they walked toward me, I knew somehow the Earth was a planet of water and land, circling the sun accompanied by other planets. It was if their knowledge emanated from their presence. Then when they all kneeled in front of me, then slowly looked up, I knew everything! All the knowledge of celestial mechanics, all about galaxies, quasars, how stars are born and die, you name it! And they hadn't signaled one voice to me yet, but being within their close proximity gave off this incredible knowledge!

"I was ready to believe that death had come to pass while I was sleeping in the desert, and I was in heaven! But my awareness discarded that thought immediately, as though a powerful new strength of reality had flourished within my soul as well!"

"Calm yourself, old man," David said. "What you're talking about is the most incredible event in history. So don't get so spastic and lose it, because I want to know the rest of your experience when you first met them. Have a seat, and chill out for a moment, then please continue."

Sarahan smiled, realizing David was slowly starting to accept him as a friend.

CHAPTER 12

ELUCIDATE

David eased back in his chair, scratching his head. Then got up and walked around, trying to absorb everything Sarahan was telling him. He looked at the old man and asked, "So what you are saying, is that these creatures actually give off knowledge and facts without knowing it?"

"Not actually, David. I was certain they were quite aware of it. But I could also feel they could block this secretion of information with no effort whatsoever. It's just that they felt they had no need to, and they were so pleased and honored that I was not resisting their … how should I say … their 'gift' of information.

"Another astonishing feeling was, as they all gathered before me, I could actually feel a sense of apprehension, a little fear of me, before they kneeled down. They were all a little intimidated that their apprehension of me might have upset me in some way. But after a moment, they knew I was in mortal fear of them, and so they filled me with passive feelings I could never describe.

"That is when they bowed before me, and I could feel their relief, knowing I could absorb their gratitude and feel their honor by having me as their guest, with no ill intentions. Then they approached me with their proposal.

"By now I was completely in control of my thoughts, and never more sure of any decision I had to make. In fact, once I agreed to the test, I could foresee exactly what this test would consist of; no experimental tables or examining machines, only to sit, or stand, even walk around if I wished.

"Not five minutes passed, when the one leading them when I first arrived, came into the room, nodded his head and telepathically voiced to me their test was completed. I asked through my new telepathic ability exactly what did they do?

"He replied, stating a very elaborate examination, and now they must reveal to me their discovery. A few more walked up behind him, and from them all I could feel utter astonishment. He asked me to please sit down, which I did, but could already feel they had made some sort of incredible discovery about me, as well as the entire human race.

"Go on," I voiced to him.

"He replied, "'Until now, every being we have ever encountered, including ourselves, utilizes their entire mind to control every portion of their physical existence. Nevertheless, the final results from your examination cannot reveal several of the most unique mysteries we have yet to discover.

"Your race has the most powerful mind capability we have ever encountered. What eludes us is that only a small portion of your brain has, perhaps ten percent or a fraction more, contains certain type of cells that are so inconceivable, our technology cannot determine their true function.

"What this astonishing discovery has revealed is that not only do these cells radiate the highest level of intelligence potential we have ever discovered, they are slowly increasing in number as your race progresses in its stages of evolution.

"In conclusion, your brains generate only a fraction of the energy they are capable of producing. But there will come a time in the distant future when, through evolutionary growth, these cells will progress enough in your brains that you will fully utilize them. This will give each one of your kind virtually unlimited power. Power of such magnitude, that neither you nor we can even begin to comprehend.

"We can only determine that the celestial forces that govern this universe elected to withhold the full use these prevailing cells throughout the rest of your brains, as our results clearly show your mind's structural psyche could clearly endure their presence with no ill effects whatsoever.

"This would explain why your civilization has advanced much more quickly than any existing race we have yet to explore, using such a small portion of your mind.

"Since the beginning of our existence, this is the most extraordinary phenomenon ever to be discovered in the history of universal evolution. We are compelled by the laws of scientific evolution to plead for another request of you."

Sarahan stopped for a moment. David noticed a tear began to crawl down the old man's face. "Are you crying? What's wrong?" David asked.

"What I have just told you, David, took less than a second to understand, from their, and now my, telepathic ability. I already knew their request within another second. It was to return with them to their home world, as the most honorable being ever to visit their planet."

CHAPTER 13

JOURNEY

"Before we departed, we had to wait three weeks," Sarahan continued with his story, "They had summoned two more of their magnificent vessels, to stand guard over the Earth while we were away, in case any outside explorers accidentally came into our system that might wish to visit Earth. I could feel their serious concern about what could happen if another race stumbled on to what they had discovered.

"Watching the other two ships approach was absolutely breathtaking. Though they both were identical to the one I was on, seeing how they broke formation to take their designated positions was like nothing any science fiction movie today on Earth could ever illustrate.

"They moved apart with such gracefulness, then slowly faded with distance."

"Must have been pretty spectacular," David said. "From what I've seen from the outside of this one, watching two more glide in together would have been pretty incredible. I'm sorry. Go on."

Touched a little more now with David's apology for interrupting, he continued, "I felt their presence behind me, about fifty of them kneeling again before me, requesting my permission for departure. I told them I was ready, then turned back to the view of space. The stars shifted from right to left, then stopped for a moment. Those beautiful dancing colors instantly surrounded the entire vessel, and blazed to a blinding flare.

"The stars before me suddenly streaked past, moving faster with each passing moment, until they turned to one bright, steady glare ahead, as if all the stars came together as one. I thought I had reached the ultimate moment of my life, hurling through the heavens, guided by unseen angels. I would have never believed that this was only the beginning of the wonders that were to come.

"They offered to put me into hibernation, as it would take about three more weeks to reach their home planet. Not a chance, I instantly replied. I was not about to miss any of this new act of providence the Creator of all things had blessed upon me."

"That's always been my greatest dream." David said. "To soar through the galaxy watching the stars actually pass by, exploring all the wonders of the galaxy. I'll bet the farm there is beauty that can't even be imagined in the realm of this galaxy alone."

"Trust me, my dear friend. You would not lose your farm. In fact, if you had telepathic ability, I could show you things your mind could not generate enough imagination to even dream of."

"So what happened next? How did the trip go?"

"Truthfully, they were so eager to please me, constantly coming into my quarters so often to see if I needed anything, I began to get annoyed. I finally had to explain to them that our race enjoys solitude, not continuous service. Then they spent the next few days apologizing, which was even more irritating.

"Though their race is far superior to ours, what makes them even greater is their kindness, which to me far surpasses any technological advancement. And with over five thousand of them aboard, I never had to worry about being alone. I was treated well."

"Like a king, eh? Come on, you had to feel a little sense of power. It's human nature. I would have loved it."

"You would have thought so, but the incredible joy and peacefulness within by being, which actually emanated from them, was shared by all.

"But there were times I thought of home, my friends, even my camel Josh. I missed them so."

Saharan slowly looked down, as if he felt he had deserted them. But when his chin rose, he was smiling.

"What?" David asked, "What's the big smile about?"

"In due time, David. Let us just say that, once I returned, I was with them again. But not in a physical presence."

"Your village," David said. "I wonder what they thought when you didn't return from the desert the following morning."

Sarahan began to laugh.

David was grinning as he asked, "What did you do?" Sarahan's laughter rose to a roar. "Come on, out with it!"

Sarahan finally caught his breath enough to continue. "They searched for days, and could not find a trace. Finally they all had decided I was taken away by angels from the heavens; which was not far from the truth. But what really convinced them was when they watched my camel Josh dematerialize before the entire crowd!"

For the first time since the two had met, they both laughed together. When David was finally able enough to ask, "Don't tell me you brought your camel aboard!"

Sarahan was able to nod yes, and then said, "I was lonely!" That was when they both went into hysterics.

When the laughter finally subsided, David looked at Sarahan and said, "I'm hungry. What about you?"

"Famished. Let us feast, and then sleep. I will continue the tale in the morning, of what transpired once we arrived at their world. Come."

Sarahan put his arm around David, both still giggling, and they left the room.

A friendship had finally formed.

CHAPTER 14

FRIENDS

David stood in the very front of the bridge, looking out at the plane of the Milky Way, so magnificently majestic. Immense as it is, it was hard to grasp the reality that it was only a mere drop of water, splashed into a sea of endless eternity. How, he thought, could our Creator sculpt such ceaseless miracles, from here to the edge of the universe, if an edge did exist.

"Good morning, David. Slept well, I trust?" Sarahan asked, as he slowly walked into the huge domed room.

"Very well. The best since I left Earth. And you, my friend?"

It enriched Sarahan's heart, to hear David refer to him now as a friend, which he tried so hard to be to him since his arrival.

"Always. Especially after our little party, so to speak. We must do that more often. I really had forgotten the joy of good ole human laughter. It was so refreshing."

"Had a great time, myself. Speaking of enjoying yourself, I can't help but ask, how do you entertain yourself, all alone out here? Don't you ever get bored? I mean, as enormous as this ship is, after twenty-five hundred years, you must know every part of it, from stem to stern."

"Bored?" Sarahan replied, as another smile emerged. "Hardly, David. There are many forms of entertainment. I have read every book ever written. Not the original prints, either, but I can scan every library, bookstore or any institution containing every form of literature ever created. Every book I scan, I know its full contents in the wink of an eye. I have seen every movie ever produced … well, most of them. My favorite movie, 'Ben-Hur', I have watched at least four or five hundred times."

David tilted his head a little with a look of confusion, and then said, "Four or five hundred times? What kind of moron would watch a movie that many times? I don't call that boring. That's sounds more like an obsessed fanatic!"

Sarahan took a couple of steps back, smiling, put his hands on his hips, and said, "Really, my friend! And how many times have you watched the movie, 2010?"

A grin slowly grew across David's face, and he was barely able to say, while beginning to laugh, "Not five hundred times ... I think. Anyway, don't change the subject ... Wait a minute! How did you know that?"

Sarahan was laughing now, and answered, "Lucky guess, perhaps?"

"Right, you space peeping Tom! That movie was about the human race advancing to space exploration, which was my very goal. Not about a bunch of nit-wits slaying each other on buggies! Unless, of course, you have a thing about guys with their short skirts blowing upward by the wind!"

Both were just showing each other the little boy that exists in all men with each statement, laughing out loud now.

Sarahan finally caught his breath enough to say, "Buggies? Not an accurate description, my friend. Chariots, they rode."

David, now with tears of laughter pouring down his face, said, "I knew that!"

"How do you think this magnificent vessel got its name, *Merkava*? I liked that movie so much, and after over two thousand years, it never occurred to me that this mighty ship was without a name. So I named it 'Big Chariot'. However, in my native language, 'Big Chariot' is translated into as *Merkava*, which I preferred, since it had a beautiful 'ring' to it."

David was doubled over by now, sitting on the floor laughing hysterically, trying to say something, but could not catch his breath long enough for it to come out.

Sarahan, still giggling, shouted, "What? I think it is a very appropriate and lovely name. After all, is this not a 'Big Chariot' of a sort?"

David was now on his knees, still out of control with laughter, with the side of his face pressed to the floor, he finally was able to say, "Then, where's your skirt?" Then he busted out screaming again, out of control.

That one caught the old man completely off guard, and as he could feel the uncontrollable surge of laughing hysterical charging its way from out of him with greater force this time, just before he erupted out of control he was quickly able to spurt out, "If you had any intelligence whatsoever, my friend, you would have assumed by now where all my skirts are. At the cleaners!"

He then dropped to the floor, also no longer in control, so overpowered by laughter.

After they were able to regain a little more control, they both sat next to each other on the first step that lead to the pilot chair, wiping their faces off.

"Okay. Okay. Let's call it a draw." David said, still stuttering a bit, then resumed their original conversation.

"How can you do all of these things. Do you actually pilot this vessel somehow all by yourself?"

"Yes, and much more. You will know when I finish telling you why they brought me to their world. But for now, back to how I keep myself occupied.

"Though my first duty is protecting our world, which in itself gets quite eventful at times, there is one thing I do that is quite unique, to put it mildly. Fascinating, would be a better description."

Sarahan walked up and stood next to David, looking along with him to the vastness of the galaxy.

"Look out there. What do you see? Magnificence. And only a tiny part of it. Our race is now upon the threshold of breaking the confinement of Earth, and will one day soon have the power to begin to reach out and leave the Earth, and will know the gratification of exploring it."

David looked down at him. "That, I'm sure, won't be for a while yet. Many of the scientists and colleagues I worked with still want to just explore space from their cozy little spots on Earth, where it's nice and safe."

Sarahan looked back at him, and said, "You are quite mistaken, my friend."

"What do you mean?" David asked.

"I mean, David, it has already begun. Look at the glass in front of you. What do you see standing between the glass, and the beauty of the universe?"

David looked out again. After a moment, he said, "Only my reflection. Nothing more."

"Much more. It is your reflection I speak of. And how is it your reflection is there? What action did you undertake to see your reflection here, instead of from a window on Earth?"

David smiled. "I see your point. But remember, the other reflection in the glass got out here first."

"Yes, my friend, but it was not by an action that I performed on my own. I had plenty of help. You did not. You are the first in the history of our existence, David, to reach out for the stars, and make it this far, all on your own. Others will follow one day, but you are the first of us, to leave the earth for the stars."

"I'll take that as a compliment. Thank you. It was a lot of research and work, not to mention humiliation. The main ingredient was motivation, wanting not just to look at it from a stationary observation platform, but to be within the splendor of this vast magnificence.

"Though I did not make it that far, for which I thank you again, seeing as how you kept me from living in an alien zoo, I found something much greater than the physical beauty of the stars and galaxies."

"What else did you find, my good friend?" Sarahan asked.

"A friend. Something I thought I would never find again. And you have restored my faith a little in humanity. I no longer hate my fellow humankind, I just pity them now."

"Thank you, David. I could feel your goodness within you from the moment we met, which was a little hard to see through all that needless hate. But with all my abilities, I could not fulfill the need for a friend.

"Even with all the entertainment features at my disposal, nothing can replace the need for company, but even more now, I am blessed with a friend. Thank you, David."

"No charge, Sarahan. Because I too, am blessed with a friend, somebody to talk to, and especially someone to laugh with. I can't remember the last time I laughed that hard. It was really nice.

"But you speak of entertainment features at your disposal. If you have read every book, watched every movie or television program from Earth, after two-and-a-half millennia, didn't things get a little redundant after a while? I mean you can only play so many games of solitaire, before tossing the deck behind you."

"True, David. But as I mentioned before, there is one thing I do that no one could ever get bored with."

"Fascinating, you called it. Tell me about it."

"I'll do better than that, as describing it will never do it justice. I'll let you experience it. That is, if you think you can handle it!"

David detected a note of funny sarcasm, as if Sarahan was challenging him. "If you can do it, I know I can. You're on! So what is it that we do, mister tough guy?"

"Tough guy," Sarahan repeated, nodding his head and smiling. "I like that. Let us see if I can refer to you as the same. Climb up into the chair."

"Your chair? Up there? Tell me first what it is we're going to do, then I'll sit up there."

"I will let you know once you're up there, or are you too afraid to try something an old man can do, but you cannot?" Sarahan knew that would get to him.

David looked up to the chair, then turned back to Sarahan smiling, and said, "You're kidding, right? Okay! As they say on Earth these days, we'll see 'who's the man'!"

David walked up the small set of stairs with a touch of macho, turned and sat down. He locked his hands behind is head, looked down at Sarahan's grinning face, then shouted, "Okay, let's see what you got! What is it that you think I can't handle, big guy? What are we going to do?"

Sarahan was laughing by now. He folded his arms, turned and walked directly to the very front of the dome's edge looking straight ahead, and replied, "We cruise."

CHAPTER 15

CRUISE

The ship began to turn, then slowly moved toward the inner part of the solar system. David noticed Jupiter pass them from a distance, realizing they were headed for the asteroid belt. David realized the view from the chair was a little better than from the floor. He was leaning forward, twisting his head from side to side, waiting to see what Sarahan had up his sleeve.

Sarahan remained in his same posture, standing at the very front of the dome looking outward, with his arms folded, in his semi-conscious state. But with a hidden grin he did not want David to see.

The asteroids began to scatter away to clear their path as the ship approached, and within seconds they were clear.

"Mister driver, you forgot to furnish me with Kleenex for when I start crying! Those asteroids really scared me, and I'm going to wet my pants!" David was trying his best to agitate his concentration. Sarahan ignored his every remark.

The ship began to roll to the left, accelerating at the same time. Shifting back to even keel, a tiny orange ball popped into view. David heard the engines roar with more intensity, looked up and the planet Mars almost filled the entire view of the dome.

"Playing chicken with planets, are we?" David remarked.

Sarahan still did not budge, but his grin slowly grew into a smile. He banked the ship hard left, while David thought he was going to assume an orbital position, but the ship was gaining more speed.

Suddenly the ship plunged downward, accelerating even more. They pierced the thin atmosphere in less then a second, with the red surface expanding into their view so quickly, David's fingers started sinking into the chairs arms. Now the vertigo began to swim into his mind, frightening him more as the surface was closing fast.

"Sarahan, just what exactly are you doing?" David quickly asked, now with a tone of concern.

The ship was moving with incredible haste. David was so scared by now, he did not notice that its descending trajectory had pulled up a little, just before he was sure of a definite impact.

But the angle of descent still was not level enough to avoid crashing into what appeared to be the greatest mountain David had ever seen, coming into view so fast, his eyes widened to their full extent.

"Okay! I give! Big mountain! Sarahan! Hel-lo down there!" David kept shouting. Sarahan still retained his composure, but a full smile had now crossed his face.

David turned his head with his eyes shut, just when the ship jerked upward, barely skimming up and over the surface walls of Olympus Mons.

Once the ship streaked over the summit of the solar system's greatest mountain, Sarahan boosted the velocity again, soaring with such intensity the red sanded surface fifteen miles below blew a dust cloud wake in their rear path.

David was so scared, but much too mesmerized to feel any fear. He was never more captivated in his life, looking back to watch the highest known mountain fade into the horizon.

"Hey, Sunday driver!" David yelled down to him. "How about an encore! Or is that the best you can do?"

"I'm sorry David. The engines are much too loud. I cannot hear a word you are saying."

Shivering with delight, David pulled himself up, and made his way down the steps to confront the old man. Once he reached the floor, he walked past Sarahan, over to the ledge where the glass began, and slowly peered over the ridge. He saw the grayish-red thin clouds swiftly flashing past them.

Still breathing hard from vertigo, he turned back to Sarahan.

"That was incredible! Come on, turn back around and do that again!

"Calm down, David. The cruise is not over yet. It gets better!"

"Better? What next?" David was too excited to notice that Sarahan was not looking directly at him, but just over his shoulder. When another smile from Sarahan instantly grew, David quickly swung around.

Before he could completely focus on the massive dark basin that took up the entire surface and their view ahead, the ship again began to plunge downward again, rapidly picking up speed.

David froze for a moment, stiff with vertigo again, watching this titanic vast gorge expand immeasurably before him, and they were closing down on it rapidly. The growl of the engines increased even more, watching gigantic mountains and rock structures within the abyss of this incredible canyon, grow quickly into view with the momentum of the ship's velocity as they approached.

Holding his hands in the air, Sarahan burst aloud, "Behold, my friend, David Adams! The Valley of the Mariners!"

David staggered back to the edge of the glass again, then began yelling, "Oh my God! Sarahan! You're really going to kill us! Pull up! Pull up!"

But the engine's roar was too loud to hear anything by now. David ran back up the steps and jumped into the chair, turned and screamed as loud as he could, "All right, old man! Let's rock! Yaa-hoo.........!!!

And the ship swiftly dropped below the top edge of the surface, plunging down into the enormous canyon at incredible speed.

Sarahan stood before this vision with absolute bliss, as he had many times before. Watching the mountains and structures of rock that were taller than skyscrapers, blaze past so gracefully, knowing if these beautiful wonders had eyes, they would never be aware of the ship's presence.

Deeper into the endless ravine they soared, just above the valley's floor. Every mountain, cavern or rock formation that jumped in their path, the mighty ship maneuvered swiftly around. Upwards they climbed; sideways they turned, rolling full circle at times.

David had tears flowing down his face, he was so caught up in the moment, knowing he was living the greatest moment of his life. He yelled down again, "Ride'em, cowboy! Is that the best you've got?"

Sarahan looked up at him and smiled, turned back forward and banked the ship hard right, increasing the speed even more toward the valley's massive cliffs, much too far below the surface to see the top.

The canyon's cliff walls rapidly were rapidly getting closer. Sarahan began banking the ship more to the right. The *Merkava* kept rolling over until the vessel's bottom side skimmed just above the valley wall.

He kept the ship in that position while they glided alongside the valley barrier for hundreds of miles, soaring with the same tremendous speed, still skimming just above the cliff walls with the ship's bottom side.

David was hanging on for dear life, but loving every moment. He knew if he wasn't going to survive, that this would be the best way to go.

Sarahan then rolled the ship over to a level plane, back to the valley's center, and plunged the ship down further into its enormous depths, rolling and turning, barely missing each crevasse and ravine that came before them.

Sarahan compensated each move with ideal precision, to any massive structures that stood in their way, yielding to none.

The *Merkava* climbed up to the surface level, at times, to take in a quick view of the valley's majestic expansion that stretched on to forever, then plummeted swiftly back down into the endless realm, to glide again with blinding speed over, up and around the graceful sculptures carved by the creation of the canyon floor.

Sarahan finally sensed the valley's end approaching. Waiting until the last possible moment, he watched as the massive cliff wall came rapidly before his eyes.

At the last possible second, the ship's mighty engines surged one more inconceivable explosion of power, and hurled the ship vertically upward along the cliff wall at immeasurable momentum, taking them up and out of the great valley's vastness.

He turned and watched as the grandest of all wonders faded swiftly into the distance, and whispered, "Until we meet again, old friend."

Then turned to watch the beauty of the heavens burst into view, splitting through the Martian atmosphere, back again to the void of space.

CHAPTER 16

RECOVERY

Hovering over the rings of Saturn, the ship gleamed from the extravagant planet's reflection. Another grand wonder, Sarahan thought, still standing in very front of the domed glass.

He heard a soft moaning from behind. David had fallen asleep just after they left Mars, perhaps from the overwhelming excitement he endured during the 'cruise'.

In a loud tone, Sarahan shouted his name.

David's eyes slowly opened, pulled his head up, then started looking in every direction he could see. Then he suddenly jumped out of the chair and starting shouting, "Okay! I'm good! What's next?"

Then he ran down the steps, and started looking in every direction. "Where are we?" he asked.

By now Sarahan was laughing so hard, he had to lean against the glass of the dome to support him, and between his breaths he was able to say, "I did not think the journey we just took would be too much for you to handle, and cannot believe you passed out." David's face began to form a grin that stretched into a smile.

"I didn't pass out, if that's what you think. I was just resting my eyes. "

"If you did not pass out, then I return the question. Where are we?"

David looked out of the dome, then turned back and said, "We, ah, are, let's see, um..... orbiting Saturn?"

A few moments later, they both were laughing so hard together again, as they did the night before, they could no longer stand, and slid down where they stood.

Both were lying on the floor in hysterics. Sarahan was able to catch his breath long enough to say, "So who is the man?" Then he held his stomach and rolled to his side, now in tears.

David was able to sit up just long enough to say, "It's not 'who is the man', it's 'who's the man', which is me! You're just an insane pilot!" Then he laid back down and rolled over on his side, in tears also.

Finally the laughing subsided again enough, and they both just laid there catching their breaths. Sarahan suggested, "Let us have a drink!" David responded with, "I'll have a double! No, after a ride like that, make that a triple!"

Sarahan began to laugh again, and said, "Whatever you wish, my friend. I think you have earned it this day!"

"This day, my friend, was the most incredible day of my life! I still can't believe it. What a rush!"

They got to their feet, stumbling a little, David putting his arm around the old man this time, and headed for what David now designated 'the lounge'.

Sarahan had a couple of glasses of wine, but David drank so much Scotch, he eventually passed out. Knowing what David must have gone through during the 'cruise', he let him drink as much as he wanted, then carried David to his sleeping quarters.

Before leaving David's room, he looked back at David, lying on his bed in a drunken slumber, feeling blessed that he had found a friend as good natured as David.

Returning to his own quarters, he tried to remember the last time he and a friend laughed together like that, and could not. It was then he realized that David was turning out to be probably the best friend he will ever have. Especially since he had accepted the responsibility of running the vessel that protected his race, thinking he would always be alone.

But his time for finally retiring was growing near, even before he had met David. After all, over twenty-five hundred years of service to his people was more than enough.

Not to mention he was ready to move on, to new things, even greater wonders and beauty, other than the magnificence he had discovered here within his own star system.

Yes, it was time. He then began to consider a proposal he could make to the beings which honored and awarded him with this magnificent responsibility.

Yes, it would work. He left his room, heading for the bridge. It was time to summon them.

CHAPTER 17

REVEAL

David staggered into the bridge, holding his head in severe pain. Sarahan was watching him from above, relaxing in his chair, smiling at his new friend.

"Do you have a headache for some reason, David? I cannot imagine why."

"Headache is not the word for it," David groaned, "More like an axe stuck between my ears. That was way too much liquor. And it's all your fault, since I needed to drink after the 'cruise' you brought me on. It feels like we actually did hit that mountain."

Walking down the steps, Sarahan couldn't resist saying, "Care to visit the volcanoes on Jupiter's moon Io today?"

"That's not funny." David moaned.

Sarahan walked up to him and said, "Look at me."

David turned to face him. Sarahan raised his hands and touched each side of his temples. A tiny flash of those same dancing colors sparked from his fingertips.

"Thanks," David said, as Sarahan pulled his hands away. "What did you do, doc? My headache is gone. You'll come in handy next hangover."

"Performed only a small effort of the capabilities I have. Which reminds me, before we took our little journey around Mars, I told you I would finish the tale about my journey to their home world. It would answer a lot of your questions, especially about the abilities I possess. Would you care to hear the rest?"

"Yes, I really would. Want to grab some breakfast while you tell me?"

"Of course. The view from the dining area is actually my favorite."

Leaving the bridge, Sarahan began, "Now were did I leave off?"

"You were telling me how freaked out your village was when you beamed up your camel. Whatever happened to, what was his name, Josh?"

Sarahan leisurely smiled. "I kept him onboard for a while, but from the moment he arrived, I could sense extreme confusion within him. He felt so lost, he actually did not recognize me. After a while, I returned him to Morocco, within a herd of camels owned by a rich man. I visited him sometimes, until he passed on. I still think of him on occasions.

"But the day he passed away, was the day I knew I would always be alone up here."

"Sounds as if you and he were pretty close. Sorry you had to lose him. But if it makes you feel any better, after my wife left me for someone I thought was my 'colleague' … someone who had convinced her I was nuts, along with the rest of my 'fellow scientists' and so-called friends, was the day I knew I would always be alone. And I was on Earth."

"If you do not mind me asking, where on Earth did you live?" Sarahan asked.

"I grew up in Cincinnati, Ohio. Went to grade school there, high school, and graduated from the University of Cincinnati, with a Master's degree in astrophysics. After college, I jumped around the United States, looking for work at any company that was interested in the study of nuclear physics. Since part of the world was utilizing the energy produced by the process of nuclear fission, my interest was always to take the next step, and reach for the knowledge and control of nuclear fusion.

"But everywhere I applied, no one had any real curiosity in nuclear fusion, because they all assumed it was too complicated to figure out, much less even attempting to harness its control, and everyone thought of me as just a dreamer of some kind. The ironic part they always danced around, was that I couldn't convince anybody that nuclear fusion was the simplest atomic physical process in the universe! Two hydrogen atoms colliding at the speed of light, creating a nuclear force that stuck them together to make a helium atom!

"I got so angry at one interview, I got up and screamed to them that what did they think made the sun shine, then walked out.

"I realized then that the only way to make it happen, was to just accomplish it on my own. It cost me everything, even my marriage.

"Yet, thinking back, I suppose I can't really blame her. I was always tied up in my research, hardly ever at home. And when I was home, I was in my lab most of the time. It was if I really didn't know her."

They arrived at the room where they always ate, got their meals and sat by the viewing port.

David went on, "The only reward I wanted from my discovery was distance. Not just from those who rejected my ideas, but from everyone. Guess that's how I ended up here."   "I'm sorry your friends at your work did not believe in you. But on the other hand, if they did, I would not have a friend here now."

David smiled, and said, "Thanks. I got the better of the two choices. You're a pretty good egg, Sarahan. So finish telling me about your trip to their world. Before you told me about Josh, you were saying that during the voyage they had bothered you, and you finally had to tell them to back off.

"But you went on to say how kind they were, which was the better part of the trip. So tell me the rest."

Sarahan had finished eating, stood up and went to one of the lounge chairs by the viewing window. "Finally we had arrived. The approach to their planet was glorious. Hundreds of their vessels surrounded us to escort our ship in.

"They had two suns, a double star. One was larger than the other, but they both were absolutely beautiful. The larger one gave off a bluish haze, the smaller almost a dark orange.

"They orbited each other, as all objects do under the laws of celestial mechanics. However, there was a unique phenomenon occurring between them. An energy stream of such intensity flowed out from the smaller orange star, from the pull of the larger blue one. This energy band glistened with exquisite beauty, continuously fluctuating with majestic splendor.

"The two stars would push and pull from each other, which is common in the existence of double stars, due to the gravity which binds them together.

"Whenever the 'push' cycle would begin, the energy band between them would radiate so powerfully, it would cast off an incredible cascade of its excess energy, occurring from the squeeze process taking place. The more the two stars pushed each end of this energy band together, the greater amount of excess energy would be released into space."

David was fascinated. "Makes sense. If you push any object hard enough together, it will burst. Go on."

"Exactly, David. After millions of years of their evolution, they learned how to harness and utilize this endless flow of energy."

"Once they learned to control it, it revolutionized their civilization. They became the most powerful race known, as this energy was used to control their entire way of life. It is the same energy you have seen emerge from this vessel, those brilliant dancing colors.

"But they had to build massive machinery throughout their history to take command its energy. That is why this vessel, along with the rest of their fleet, is so incredibly large. This ship, my friend, is one massive conductor to sustain and control that energy."

Sarahan hesitated for a moment, took a deep breath, and said, "It is also the primary reason they asked me to visit their world."

CHAPTER 18

CONCEAL

David thought for a moment, then asked, "Does it have anything to do with why it takes hundreds of them to man this ship, but just one human?"

"Precisely, my friend."

"Why? What makes us so unique compared to them?"

"Our minds. There are cells within our brains, a great deal of them, that coincide somehow with this energy field. These cells exist in the eighty to ninety percent of our brains our scientists say is not used. For some reason that only nature could answer, the portion of our minds they believe is unused, is in a sense correct.

"However, the eighty to ninety percent of the human minds contain inactive, dormant cells."

"What? Are you kidding? Why are they dormant? They must be there for some reason," David remarked.

"Correct, otherwise they would not have evolved there. But I quite assure you, all human minds contain these unused cells. To this day, my good friend, their greatest scientists and researchers have yet to discover why they exist only in our minds, and no other races, including their own."

David was even more confused now, but extremely interested, just now learning that his race was somehow special above all others.

"That's unbelievable. I sure would have never known."

"Nor would I, David. Nor would they have never known, had they not have come into our star system on an exploration mission."

"Is that when they first took you onboard, and treated you as if you some kind of god?"

Sarahan leaned forward, looked David directly in the eye, and explained, "Yes. That is why they asked me to come to their world, after completing their initial tests just after they first brought me on board.

"But they did not reveal this to me then. All they could tell at that time was that an incredible discovery had been made, and asked, that I accompany them to visit their home planet as their guest, so they could confirm their preliminary results."

He leaned back, thought for a moment, and went on, "I accepted their invitation without hesitation. Primarily due to the fact that I was exposed to something I believed was a 'gift', which increased my entire mental capacities to a degree I never even conceived possible.

"This 'gift' expanded my knowledge of celestial and universal physics to such a great extent, that even today or in the distant future, our greatest scientists will never know, the joy and peacefulness that blended within my very soul."

"And you felt no harm would come to you, when you decided to go?" David asked.

"What I could feel was absolute trust," Sarahan answered. "You might think this absurd, but it was not the trust in them that inspired me to go, it was the trust within my own instincts.

"You see, David, I knew there was something they were concealing from me. I also knew whatever it was, was for not only my benefit, but for the benefit of mankind, and the survival of our race."

"Which was?" David asked.

"Let me finish giving you the background first." Sarahan answered.

David was too fascinated to interrupt again, and replied, "Please, go on."

"After the astounding greeting they arranged when we arrived at their home planet, we disembarked and they made me very comfortable. I was taken to a room that was fit for a king, and they appointed several servants to be at my disposal. Even the meals were absolutely extravagant, prepared with food they extracted from Earth before we left.

"But the greatest accommodation was the view the room had. A view of the two stars, exchanging that glistening band of energy. A picture that will live in my mind forever.

"Every ten to twelve hours they would bring me to a facility to perform various examinations. Actually, all I did was sit in a laboratory area for ten or fifteen minutes each time, watching different colored lights on some kind of monitor. Then they would return me to my quarters.

"These tests were completed after perhaps a week. After the last analysis was completed, they advised me that they would inform me the next morning of their results.

"Five of them came to my quarters the following morning. We all sat and they told me of the results of their final assessment."

Sarahan got up, prepared a glass of Scotch, and one of wine. He placed the glass of scotch before David, and then took a sip of his wine.

"What's this for?" David asked.

"Before I tell you what they had to say, you might need that," Sarahan explained.

CHAPTER 19

TESTIMONY

Sarahan sat back down, took another swallow of wine, leaned ahead, and said, "This is how they began."

"Please accept our sincere regret and request for forgiveness for not formally introducing ourselves. The planet you are now on is called Braveria. Our race is known only to ourselves, and now to you, as the Braverians.

"However, we sense that in all probability, you could already know this, along with much more about our race. In fact, our conclusive results to our analysis do not reveal the extent of what knowledge you know about our civilization, or the level of increased mental capabilities your mind has achieved.'"

Sarahan took another sip of wine, then explained to David, "I informed them I did already know, somehow, along with a great deal more about my world and the universe than I could even begin to imagine, after they first brought me onboard their exploration vessel.

"I further explained that during that first encounter, my mental awareness was taking in so much knowledge, along with that joyful, peaceful experience at such an incredible rate, that my entire awareness was in such a state of gratitude and bliss, the last thing that would have occurred to me was for it to cease by any means.

"However, I admitted to them that after two, perhaps three hours, it suddenly discontinued. Then asked them to please explain why these inconceivable abilities emerged within me those first few hours, then the progress of my new evolution unexpectedly terminated."

Sarahan emptied the wine glass, then rose to pour another. He looked back at David and asked, "More?"

"No, thank you. Don't get gassed on me now. I want ... I need to hear the rest! So please back off a little on the booze, okay?"

"Please forgive me, David. That will not happen. I, too, need for you to fully hear this." Sarahan just didn't explain why, at least not yet, and went on.

"They hesitated momentarily to my question, then answered."

"First, we must apologize for a grave miscalculation that occurred when our vessel's personnel brought you onboard. We do know that you are quite aware of the magnitude of energy we have obtained and have learned to grasp, and where it emanates from. Also, the enormous technological endeavors we must put forth, engineering machinery of immense magnitude, of such vast massive volume, in order to sustain its control.

"What you do not know, among the countless civilizations, races or beings of every form that exist, your race, and no more than your race of humanity, which the Creator of physical laws of universal nature can only explain, can control this energy with your minds.

"Our scientists among the crew of the exploration vessel that first encountered your world, were so intrigued by the data returning from our initial probes of your planet. The information coming back was telling them the inhabitants of your world had dormant, inactive cells within your minds. That close to ninety percent of your minds contained these cells.

"But what had them absolutely awestruck , was that our instruments were telling them our energy field was in someway compatible with the organic structure of these cells. Therefore, they choose to bring one of your kind onboard for further testing, in which they elected you.

"After the first test they performed on you, it was instantly discovered that these latent cells were beginning to generate activity. These cells were increasing to radiate with life at a phenomenal rate with each passing moment. They realized it was due to your exposure to being within close proximity of our energy field by bringing you onboard.

"They immediately shielded you from this emission, which stabilized these cells from any further growth expansion. After additional testing onboard was completed, the results concluded that no harm had come to you, with the exception of your increased abilities, and we requested you to come to visit our home world as our guest.'"

Sarahan rose and returned to the window, looking again toward the vastness of space, now smiling. "But like I told you before, I knew their invitation was more than just one of a visitor, which they did not deny.

"Upon completion of their explanation of the entire picture of exactly why they asked me to come to their planet, they made me an offer that would change the entire course of my life. One that would not only greatly benefit myself, but would also facilitate a magnificent contribution to the progress and well-being of the future evolution of our race."

He then sat before David, and maintaining his smile.

"Which is why you're here now, piloting this ship. Guarding the Earth from outside interference. Right?" David asked.

"Correct, my good friend. An opportunity that I could not possibly refuse."

David thought briefly for a moment, then said, "I'm not sure I could have refused an offer like that myself. An opportunity of a lifetime. In fact, literally many lifetimes, for as long as you have lived."

Sarahan leaned forward to David. "Then, David, take them up on the same offer they intend to present to you. They will be here in a few days."

CHAPTER 20

OPPORTUNITY

"What? They're on their way here? Now? You've got to be kidding, right?" David asked with astonishment.

Sarahan still retained his proud smile. "No, my friend. I am not joking. And as we speak, they are presently en route here."

"And they intend a make me the same offer? To pilot this ship, and to guard the Earth? Why, I mean, what's the matter with you?"

"Twenty-five hundred years is a long, long time, my friend. My service to our world has given me a sense of accomplishment that none will ever know; one that I have been proud to uphold. But the time has come for someone else to replace me. I am ready to step down, to retire. I wish to move on, to experience new wonders, to see new treasures this universe has to offer."

David was so intimidated and confused; he hardly knew what to say. "But why me? I don't have one spark of the powers you have! Not to mention, they don't even know or have ever heard of me!"

"They can teach you how to utilize your unknown abilities, by slowly exposing you to this incredible energy. Once fully exposed, your capabilities will be so abundant, you will feel different about who you are, and what you are. It will be as if a new person has been fully developed into your very soul. But above all else, the peacefulness and joy your fully developed mind will be transformed into, will give you harmony that you could never have believed could exist within your spirit.

"As far as them not knowing you, they know you quite well. I have personally vouched for you."

"Personally vouched for me?" David got up and began walking in circles, with that disturbed shroud look about him.

"Look, Sarahan. I'm quite honored. But I don't know. This is a bit much. In fact, quite overwhelming." He stopped his circling rhythm, and walked to the window, gazing out to take in the view.

Turning back to Sarahan, he said, "Listen, I'm really gratified, but…"

Suddenly Sarahan sprang to his feet, and headed for the doorway. "Quickly, David, we must hurry!" Then out into the corridor he ran. David could hear the engines powering up to full strength, and was soon trailing a few steps behind him.

"Company again? Never knew or seen you move this fast before! What's happening?"

The only response the old man shouted back was, "We must hurry! Come!" Then he turned into the bridge passage and up to the very front of the dome.

The ship was already underway, heading toward Earth at unbelievable speed. The engines roared with deafening thunder, hurling up and over the asteroid belt this time, instead of plowing gracefully through as he always did before. The view of Earth was rapidly closing toward them.

David ran up behind him, looking over his shoulder. "What's going on now? Visitors again?" he quickly asked.

Still catching his breath, he said, "Somehow they got by my sensing, and are already hovering above Earth. This could only mean one race. A race that has the ability to stealth its approach. The only way to detect their presence is when they come within a certain proximity of Earth. You could be in for one incredible ride, my friend. There!"

Sarahan was pointing at two small yellowish pinpoints of light, moving away from Earth, increasing their distance with each passing moment. Sarahan blasted the engines so hard, the mighty ship was actually trembling, and came up behind the wake of both ships in under a second.

Both ships flared a reddish streak at the same time, and headed off in opposite directions.

Sarahan fired a small shot to the one that headed left, while banking the ship hard to the right, and within a few moments they were both looking directly at the stern of the other vessel.

Another small shot blazed out from Sarahan's ship, but to David's surprise, only crippling its engines. Both of the smaller ships now drifted slowly in space.

"How come you didn't annihilate them, like you always do? Showing a little mercy in your old age now?" David smiled as he asked.

"No, my friend. Both vessels have humans onboard, which I can well assure you, not under their own free will," Sarahan said, with a tone of regret.

"They were abducted?"

Sarahan nodded yes, then David asked, "So now what."

"Watch and learn, David." A crystal-like beam emanated from the ship, surrounding both of the other smaller intruders. Once the haze of the beam faded from the two ships, the colorful bands blazed together at the ship's bow and stern, and swiftly collided at the center of the ship blasting the two intruders to atoms.

CHAPTER 21

AMBUSH

"Did you just blast them along with our own people aboard out if the sky?" David asked with annoyance.

"Of course not. It was necessary to get an exact location of them onboard the two vessels first. That is why I had to disable the two ships first. Once I transferred our inhabitants, then I destroyed them."

"Transferred them where? Back to Earth?"

"They are onboard our ship, in a form of hibernation stasis, until I return them to Earth. You must really learn to calm yourself, David. Please fix yourself a drink, if you must."

"Our ship." Had a nice ring to it, David thought.

"I'll pass on the drink. So when are you going to send them back to Earth?"

"When I complete the analysis of them. I must be certain they were not injured or harmed in any way. But the most vital part is to remove from their minds any memory of this incident. I must also complete this as quickly as possible. The faster I get them back, the less time will pass that their instincts will have to account for. "

Made sense, David thought. He then asked, "How do you know where to return them to?"

"They will tell me. Their memory of their last awareness, that is. It will only take another moment."

Still somewhat confused, David turned and looked around the entire domed bridge area, or "command center," as Sarahan called it.

After skimming his eyes over every part of the bridge, he asked, "Just out of curiosity, how do you control anything on this ship? I mean, every since I've been aboard, I haven't seen one instrument or control panel. Do you actually operate everything on this ship with your mind?"

"Everything. The energy field that envelops this entire vessel is part of my consciousness. There. The analysis is complete. They are all on their way back to Earth now, each at the exact location they were removed from. None of which have any memory of their abduction. " He began walking up the little stairway. "To further answer your question, an extension of my mind, would be another way of describing how I control this ship."

Once he got to the top, he turned around and sat, in what Sarahan referred to as the "command chair"; or "big chair", as David called it. "You see, I can reach…," His voice unexpectedly stopped. He quickly leaned forward, then began twisting his head from side to side.

"What is it?" David asked while turning to look outward.

Sarahan suddenly shouted, "David!"

David yanked himself around. Sarahan was already heading down the steps, when the entire vessel was struck by something that shook it with one mighty jolt forward.

Sarahan held onto the right guard rail of the steps with both hands, watching David fall face down to the floor. Another powerful jerk instantly followed, as if something slammed into them from behind. Sarahan could not hold on this time, and was thrown backward, striking the back of his head against the steps.

"What's happening?" David screamed, and hit the floor even harder this time. Sarahan did not answer. David tried to look upward at him, when a third incredible jolt slammed him back down, so hard he saw a blue flash from the impact of his forehead striking the floor.

Barely conscious, David kept shouting his name, but no answer. Two more powerful jolts shook the ship again, hurtling David just above the floor each time. Five more followed. David was completely helpless, bouncing off the floor each time the mighty ship was struck.

Finally the strikes ceased. Fear had him frozen stiff, but the startling silence petrified David even more, wondering when the next earth-shattering strike was going to occur. Calling out Sarahan's name again, quiet stillness was his only response.

Being so overwhelmed and preoccupied with what was happening to the ship, he did not realize that something else was starting to emerge in the back of his mind.

All his fears, that had him completely immobilized, were slowly beginning to fade, replaced by a sense of forward motion to take action. Instincts brought him to his feet before he realized he was standing.

He glanced down and saw Sarahan lying in a crumpled position at the bottom of the steps, completely motionless. He rushed over to his side, and knelt down beside him. The old man was completely unconscious.

"Sarahan?" David said, while he was leaning him upward. He noticed a little blood was dripping from the back of his head, which had apparently struck one of the steps during the attack.

Attack? How did he know that? But he quickly put the thought aside, since treating Sarahan's wound took priority. David pulled his shirt off and pressed it against the injury. Though the old man was unconscious, he knew there was no concussion, and realized he would be out cold for almost an hour, and would wake up with one serious headache. He slowly laid him back down, resting the back of his head onto his shirt. The bleeding began to subside.

He jumped up and turned to head for the front of the dome, but hesitated. Looking back down at Sarahan, he thought again. How was he so sure there was no concussion, and that he was okay and would regain consciousness in less than an hour? He knew it was more than just a lucky guess. David's instincts were positive it was true.

But his awareness immediately snatched him away. He turned to see those same types of ships, hundreds of them, hovering about a thousand meters outside the bridge dome, except these were much larger than the two Sarahan destroyed. David's new sentient thoughts told him their ship was not without power, but adrift with no control.

His new confident awareness he now possessed, also told him their ship was completely surrounded.

CHAPTER 22

DESPERATION

David was awestruck by the sight he was looking at; an armada of alien vessels, holding their ship in check. He walked back over to Sarahan, still lying on the floor, and said, "Would you please wake up? We cannot wait an hour. We have company, and I'm not much of a host."

To his surprise, Sarahan slowly opened his eyes, winked a couple of times, and then raised the upper-half of his body. Holding the back of his head, he dropped into that semi-conscious state for about two seconds, opened his eyes again, and then stood up.

"Welcome to the party. Are you all right?" David asked. Sarahan only glanced at him for a moment, with no response, then walked to the face of the dome.

Without looking back, he said, "I am uninjured now."

He walked over to David, then took a step back and looked about his head and the rest of his body. "I do not think removing your shirt was enough to scare them away." Then he stepped closer, looking straight into David's eyes. "Are you all right, my friend?"

"Very funny. And yes, I'm fine. In fact, better than fine. But never mind that right now. What about our friends out there?"

Sarahan didn't seem interested in their assailants, and just kept staring at David.

"Why are you looking at me like that? Hello?"

After a few moments, Sarahan turned back to look at the fleet of ships, still motionless. Finally he said, "We were lucky. Our ship sustained no harmful damage. However, we are adrift. Though I am attempting to communicate with them, they are not responding.

"Our visitors believe our vessel is damaged, and think we do not have control. But they are uncertain, and are observing us to see if we are absolutely immobilized. For the present, we must continue to let them believe that."

"Why? I thought by now you would have blown them all to kingdom-come. If we're not really damaged, why are we still drifting?"

"It will only take a few moments for me to regain control. But the instant they detect that, they could resume their attack. We must proceed with extreme caution."

David looked out at the hovering fleet, then turned back to Sarahan. "I thought this ship was invincible."

"No vessel is indestructible, David, not even this one. If it were only a few of them, there would be no problem. But hundreds of their ships surround us. Their combined firepower, which they no doubt utilized upon us, could possibly destroy this vessel. That cannot happen."

"Then why didn't the Braverians take this into account?" David asked. "I mean, surely they must have known something like this could happen. Right?"

"Not their entire fleet, my friend. This is truly remarkable. Something must have happened to them; a planetary catastrophe perhaps, that would motivate them to this level of desperation."

"To what level of desperation?"

"Desperate enough to dispatch what appears to be the majority, if not all … every vessel they have, to capture one or several members of our race."

Sarahan looked over to David, and noticed he was beginning to stagger. Suddenly, David dropped to his knees. Holding his hands over his face, he began to fall to his left side. Sarahan was able to catch him in time, and helped him over to the first step that led up to the chair.

"David, are you injured?" David just sat on the step, not moving his hands away from his eyes.

Sarahan shouted his name this time. David finally dropped his hands, but did not raise his head to look up.

"Yes, I see what you're saying," David said. "Also, desperate enough to leave their home planet Caperia practically defenseless. And not knowing what guards the Earth, it seems as though they used the first two vessels to lure us out. Once they knew we were completely focused them, they slipped up behind us. And now, the next move is up to them."

Sarahan was completely startled. He stared at David with amazement, then muttered, "Correct, my friend." He stepped back, still looking down at David. "David, when I was unconscious, did anything unusual happen? I mean, to you?"

David continued to stare aimlessly forward. "I'm not sure. But the minute you fell unconscious, I felt a surge of new awareness slowly begin to creep its way into me."

He began to leisurely raise his head to face Sarahan, and went on. "Things I couldn't possibly know, yet did." He took a couple of deep breaths, and stood up.

Walking to the front of the dome looking outward, David watched all the alien ships slowly bobbing around them. Sarahan walked up behind him. "Please continue, David. It is important that I know."

"I was so scared, thinking we were going to die for sure. All of a sudden the panic just disappeared, a few seconds after you were knocked out. It's hard to explain." He turned and looked at Sarahan's bewildered face.

"My instincts began to tell me that it was not something, like an asteroid or a misguided moon that hit us. Somehow I knew it was close to a thousand ships, firing at us from astern. When they stopped firing, I knew; actually I could feel, they were surrounding us. My attention immediately shifted to you. After laying you back down with my shirt behind your head, not only did I know you were okay, but you would also be awake in about an hour. Five minutes later, when I was actually talking to myself, I asked you to please wake up, your eyes opened!

"Since then I keep getting these powerful surges of new information in my mind that just pop in out of nowhere! What's happening to me? It feels like what occurred to you when the Braverians first brought you on board, is happening to me. You see, here it comes again!"

Sarahan helped David to the floor, and sat him in an upright position. David's hands were back over his face again.

"That is exactly was is happening, my friend. In addition to what you have just told me, there is also no way you could known our friends outside are Caperians."

Hesitating for a moment, Sarahan looked back out at the Caperian fleet, and continued, "The moment I first brought you onboard, I shielded the energy field from your presence. When I received the blow that rendered me unconscious, my screen of protection from it dropped as well."

David's hands slid down, and he looked up to Sarahan. "So now what? Can you stop it somehow?"

"I'm afraid not, David. The energy field activated your mind's dormant cells, let's just say, to the point of no return. Welcome, my friend, to a whole new universe."

David stood up, giving the impression of being in charge of himself now. "Thank you. There is no way anyone could ever pass on this new feeling of joy and peace that comes with it. Let's just hope we live long enough to both enjoy it together."

He looked back out to the massive fleet, and said, "Because from what my instincts are telling me now, our friends out there are convinced that we are no longer a threat. Hang on!"

CHAPTER 23

SURPASS

Swiftly the bombardment came. The enormous vessel began to spin, end over end, shaking like an earthquake. Sarahan and David watched the view of the star field sail by, which almost seemed like a streaking blurriness from the dome.

"Any suggestions?" David shouted.

Sarahan quickly answered, "At this point, only one, my friend."

David saw the dancing colors begin to flare. The ship stopped its tumbling, and the view of the sun appeared in the middle of the dome's view. The roar of the engines blasted with incredible thunder, hurling them away at tremendous speed, while firing astern upon the pursuing fleet.

Sarahan blasted to pieces the trailing ships that fell in their wake, but the majority of their fleet remained within range.

The view of the sun was growing larger by the moment. Sarahan turned quickly to David, and shouted, "I cannot take them all. The only chance we have is if we both combine our powers, and bear together our full strength at their leading ships."

"How?" David hollered back. "I don't know how to do that!"

Sarahan grabbed him with both hands, yanked him to within inches of his face, and into his eyes he telepathically sent, "Perceive me, David! You must concentrate your entire focus into one focal point stream. I will then signal you, and you must fully release the energy field's full power graced within you, into one swift surge of thought, keyed onto the destruction of their vessels. It will take all the power of concentration you have, but it will at least give us a fighting chance to save our race, and possibly ourselves as well!"

David was mesmerized, almost to the point of shock, of how his mind absorbed the telepathic message inside of a second. Sarahan jumped back a step when David signaled his powerful telepathic answer, "Then what are we waiting for?"

It took Sarahan a moment to gather his thoughts back together, and smiled at David with envy. Speaking with his voice now, he said, "You learn quickly, my good friend." David returned his smile, and both turned to watch the sun climb further into their view.

With both of their minds linked as one, Sarahan passed another message to him. "I'm sure you have already read and know my strategy with this plan."

David returned the answer immediately, "Yes. Let me know when you're ready."

The sun now filled the entire view of the dome. The enormous ship's bow began to pull upward, skimming just above the mighty star's atmosphere. Sarahan soared the ship just over the top of the sun's corona, then down the backside, boosting the engines with greater speed. The trailing fleet no longer had them in sight.

Once they reached the sun's underside hemisphere, Sarahan rolled the ship over on its back, with the sun's surface now above them. The pursuing attackers were now above them, all heading for the sun's topside, chasing their ship in the direction they had last seen them.

"Now, David!" With their combined effort, the energy field flared the dancing colors with blinding intensity from both ends of the ship, and rolled to the middle in less than a second. Then collided with such brilliant might, the tremendous thunder of firepower was something neither David nor Sarahan had ever seen.

Hundreds of the leading Caperian vessels were instantly blown to pieces. The colorful bolts springing from their ship began to work their way toward the middle of the fleet, blasting them apart one by one.

The remainder of the fleet finally acquired a fix on their new position. Sarahan and David began to feel thousands of impact jolts, shaking the mighty ship to an on and off course, so outnumbered they could not repel them all. The shatter of the impacts finally knocked them to their feet. They could no longer focus enough to control the energy field to maintain firing.

Their ship was leisurely spinning adrift again, just outside the path of the planet Mercury's orbit.

Lying with his back flat on the floor, looking straight up, David asked, "Do you have an alternate plan B?"

The Caperian fleet stopped firing. They had them fully encircled again, hovering just a few hundred meters away.

Sarahan managed to crawl his way up to the first step, turned and sat facing outward, watching the enemy fleet that now surrounded them. "There is only one alternate left, my friend."

David was able to stagger to his feet, looking from one end to the other of the view outside the dome. "I read you. But I'm afraid we have another crisis now to consider. Several of their ships have broken off from their fleet, headed for Earth. No doubt to take captives."

Still staring with a puzzled look at the massive fleet, David asked, "Why haven't they finished us off? Wait a second, did you hear that?"

With a tone of scorn, Sarahan said, "Yes, which just answered your question. Evidently that transmission was from the commander of their fleet. He wants to capture this vessel intact. I have misguided their sensing devices to make them believe we are completely powerless, which, I'm sorry to say, is not far from the truth."

Sarahan got up and stood next to David, both with their arms folded, looking out at their enemy's countless fleet.

David pointed to the lead vessel. "That's the one were message came from."

Sarahan turned to David, now with a full smile. "We do have sufficient power enough for one last engagement. However, you already know we will not survive."

David glanced to his mentor, and said, "You already know my answer."

"And mine as well, my good friend. Let us respond to him. Then … how do you say … 'Let us rock'!'"

David slowly grinned, then asked, "Don't you ever use contractions?"

"What do you mean, David?"

David's grin was now a full smile. Turning back to look one last time at the beautiful vastness of the heavens, he then replied, "Never mind. Let's give him his answer, and then, 'Let us rock!'"

CHAPTER 24

SHOWDOWN

In their own language, the Caperian communications officer shouted to the flagship's leader, "Commander, we have a response. Actually, sir, for some reason, there are two replies."

"State them!" He yelled back to him.

"The first one is quite clear. It only reads, 'Unacceptable!'"

"Fools! And the other?" he demanded.

With an odd look on his face, he replied, "I am sorry, commander. We are unable to decipher the words of the other response."

"State the words, immediately!" He screamed.

The communications officer hesitated for a moment, until he heard the commander scream, "Out with it!"

Slowly looking up to the commander with a puzzled face, he said, "Not a snowball's chance in hell, butthead?"

The commander looked over to his next in command, both with the same puzzled look as the communications officer.

Suddenly, both spun around when they heard the weapons officer yell, "Commander! They have just fired one massive burst of energy!"

The commander took one quick step toward him, and screamed, "Direction?"

The weapons officer leisurely looked up at him with a face of doom, and said, "Directly at....!

Less than a second later, the impact was so immense, not only did the commanding ship explode into millions of flaring particles, nine of the vessels which surrounded the flagship as guards nearby, met the same fate.

"If we were bowling, we just scored a strike!" David shouted to Sarahan.

The remaining Caperian vessels instantly began to fire back, spraying so many bolts of destructive energy in their direction, the view of the approaching rays turned the blackness of space to an almost completely white. All Sarahan and David could do was concentrate and focus all their abilities to return as much firepower as possible, which was decreasing by the moment.

Just before their energy field was completely exhausted, the enemy ships stopped firing. The hundreds of Caperian vessels quickly turned in the opposite direction, but remained still.

Sarahan and David were both back on the bridge's floor. So exhausted they could hardly breathe, they slowly looked up towards the enemy fleet.

David managed to raise his head enough to ask Sarahan, "What's happened? Why did they stop?"

Both struggled to their feet, hanging on to each other.

Between his hard breaths, Sarahan was able to mutter, "I am not sure, but something has happened to their vessels approaching Earth."

"Something's happened to them all right," David said. "Probably because they've vanished! I'm not reading their presence at all! Are you?"

"Nothing. It is as if they have simply faded away! But we must act. Now is the time to focus our concentration to regenerate our energy field."

David looked over to Sarahan. "We can do that?"

Sarahan looked back to him and said, "We can do that. Even more so with the two of us together. Just follow my concentration. Quickly, David!"

They merged their thoughts together. The sparkling colors began to slowly to radiate with power, dancing and flaring around every area throughout the ship. The two stood together, with their arms folded, standing in the very front of the dome. They were in that deep trance, with their eyes barely opened, David had seen Sarahan perform many times.

David felt it immensely refreshing, feeling the energy field come back to life.

All of a sudden, they both snapped out of it, feeling an immeasurable change jump into their awareness. Opening their eyes, the Caperian fleet was scattering in every direction.

David looked at Sarahan, who was staring out toward the enemy. "What are they doing now?" David quickly asked. Sarahan did not answer, but had a full smile glowing from his face.

Sarahan shouted for the first time since David met him, "Behold! The cavalry has arrived!"

David jerked his head to look outward, and had to take a couple of steps back, he was so incredibly awestruck.

Five magnificent Braverian vessels materialized just above what was left of the Caperian fleet. The fleeing Caperian ships were incinerated by the Braverian vessels in the wink of an eye, leaving only a huge cloud of space dust where they last existed.

"Ho-lee Mo-lee! Now that's what I call timing!" David exclaimed.

"Come, my friend. Your bravery through this experience is quite remarkable. And your new companions out there are presently intoxicated by your fearlessness. You are quite a celebrity to them now. Come, my very good friend! I am looking so forward to be standing next to you when they come onboard to worship you now. They will be onboard in a few moments."

"Yes, I feel it. But since you're my best friend, you get the first autograph!"

"Then I am even more honored, 'Mister' David Adams!"

With one arm around each other and laughing, they began walking toward the domed room's doorway.

As the hairline crease was opening, David said to Sarahan, "Do you think they'll make a statue of me?"

Face to face with trust and cheer, Sarahan said, "No doubt!"

CHAPTER 25

INTRODUCTION

They materialized onboard in one of the largest rooms on the *Merkava*. Hundreds of them materialized every five seconds.

David was even more mesmerized again. He could feel their incredible eagerness to meet him, and Sarahan again.

The sight of the Braverians also had David completely captivated. They were beautiful beings, blue-glossed in color, all ranging between five and six feet in height. But what fascinated David the most was their eyes. They had incredible eyes, with a blue shining color that David could only compare to those of angels.

Once the room was completely full, David could feel the leaders ordering the rest into formation, which took maybe five seconds. They all dropped to one knee, bowing their heads.

David could hear within himself, Sarahan telepathically saying to them, "Please rise, my good friends."

As they all rose up together, David already knew which one was the leading emissary. He stepped up to face Sarahan, and David could hear his voice within say, "Greetings, Sir Sarahan. We are most honored to see you once again. We came in answer to your summons, and proposal, but did not expect to be so honored to witness bravery of such magnitude from your companion."

Sarahan watched as he stepped over to face David. He bowed his head, and sent David a voice which said, "You are David Adams. We are duly honored by your presence. We welcome you, and are also honored."

David looked at Sarahan, who had another ear-to-ear smile, then nodded back in the direction of the leader standing before him. He turned his look back to their leader, and could feel the purest joy and sacred peace within this being.

David voiced to the emissary, "Thank you. I'm honored to meet you. I'm honored to meet all of you." The rest of the

Braverians began to gather around him, looking him over, but feeling their joy and pleasure to be within David's company.

Their leading emissary raised his left arm halfway to signal the others to give David a little space, then turned back to David, and voiced, "We know Sarahan has graced you with his proposal, but he has not as yet acknowledged your acceptance."

David looked again at Sarahan, who still held the same smile.

"If I accept your offer, what will become of my good friend?" David asked.

"He has elected to come with us, to accept his reward."

David looked back again to Sarahan, who had already stepped up to face David. "The reward of seeing and experiencing the true beauty and wonders of this galaxy has to offer."

"My original dream, Sarahan!" David turned and walked over to look outward to the beauty of the universe. He looked from one end of the Milky Way to the other. Then turned back around, walked up to Sarahan and with his speaking voice this time, said, "But I suppose you have more than deserved it. I want to see the beauty and countless wonders of this galaxy myself, but perhaps I need to earn it first."

David turned to face the emissary, and voiced, "Will I one day be granted this same reward, if I accept?"

The joyful answer poured into David's soul from the Braverian leader, "And much more, I quite assure you."

David glanced back to Sarahan, then back to the emissary, and said with his growing telepathic ability, "Then I accept."

An incredible blast of satisfaction and happiness bolted through David, coming from all the Braverians. When he looked to Sarahan, he noticed a few small tears was beginning to run down his cheeks.

"Are you all right?" David asked him out loud this time.

"More than you will ever hope to comprehend, my so-good friend." Then stepped closer to David, and tightly wrapped his arms around him, which David returned, hugging the old man as hard as he could.

The Braverian emissary stepped back, along with the rest of them, to give David and Sarahan room to embrace, someday wishing that he along with the rest of his race could feel the immense joy the two human beings was now sharing. But he also could sense a note of regret between the two, unable to recognize this unfamiliar emotion. He listened further as the two humans continued their conversation, hoping to at least get a hint of this confusing sense of sorrow.

When they pulled apart, David had now had tears running down his face as well.

"What will I do without you?" David asked while wiping his face.

"Do you actually think I would have let you accept this gracious offer, without their assurance that I would be able to communicate with you at anytime?" Sarahan asked with broken speech, wiping away his tears as well.

David smiled and said, "As we say on Earth now, a phone call is the next best thing to being there!"

"Except in this case you will not have to hold a telephone receiver. And do not think I will not be dropping in to check on you from time to time!" They both started laughing again, and embraced each other for one last time.

CHAPTER 26

BEGINNING

David felt an oncoming lump in his throat, watching the five Braverian ships' fade away into the gulf of space. Sarahan was with them, becoming light years apart as every moment passed. The best friend he had ever come to know.

No, he thought. I was, and still am, close friends with one of the greatest men who ever lived.

Off in distance, he turned to watch the distant blue-white dot of the Earth pass from the right side of the dome to the ship's rear. As he passed, in a soft calmness of his real voice, he said, "And none of you will never hear or know his name; and what he did for us. More times than any of you can count."

He turned back ahead to watch the stars leisurely moving toward him. Remembering all their moments together, led David to now see a new part of his life unfolding.

But knew in the back of his mind he was now in for even more of an adventure than that which he had first set out for, when he first left Earth. The new power from the ship's energy field was now fully bound up within him.

He was a little disappointed that it wasn't necessary to bring him back to the Braverian world, to become fully exposed to the energy field's captivation. But Sarahan later explained to him that from the moment Sarahan lost consciousness during the initial Caperian attack, the shielding he used to block its penetration into David, instantly vanished. The energy field immediately began to flow into David.

By the time Sarahan regained consciousness, he knew the energy field's dancing colors found an additional host to reside within, and they activated every one of his unused dormant cells. It happened slowly at first, as their own intelligence knew the new host could not absorb them all at once, and then they fed their own way into his mind at a rate slow enough to not damage him. Also, since Sarahan knew that David would be the next to guardian of the human race, he wasn't about to stop the mushrooming growth of them occurring within David's mind. When the Braverians first arrived, he knew then every inactive cell now radiated with full life.

Sitting in the command chair, he felt so proud of his new purpose in life. The most important lesson Sarahan had indirectly taught him, was to not hate his fellow man, but to respect their existence, and knew that one day they would rule the entire galaxy, and possibly in the distant future, the universe itself.

What made him feel even better, was that the Braverians awarded him a new ship, a somewhat smaller one than the vessel he and Sarahan had; which was over two thousand years old. This one was a newer version of their fleet.

Its new features included greater power, though David did not think was ever possible, compared to the one Sarahan commanded. It had the strength a hundred times greater, that not even ten Caperian fleets would dare to confront.

It also had a new capability the Braverians developed. The ability to conceal its physical presence, to become absolutely invisible when needed. This way no intruder could detect David's pursuit until the final moment of confrontation, which even a fleet of thousands of ships, would be no match for.

He walked down the stairs, and up to the face of the new domed glass. He looked into the direction of where the Braverian vessels faded away, still thinking of the old man. In a way, it didn't seem fair. The one person in his life he got the closest to, had to leave it; but for good reasons, knowing Sarahan had well deserved to see the inconceivable beauties this galaxy alone had to offer.

This brought the cheer back into his heart, especially knowing that anytime he wished to talk to him, all he had to do was point his new powerful thoughts out to the universe, and Sarahan would answer. Not to mention that he would be stopping in occasionally to visit.

But for the time between those distant telepathic conversations, and his occasional visits, he would be alone. However, with the entertainment features that Sarahan introduced him to, he knew boredom would never overcome him. Especially one.

He glanced off to the left, and saw the distant red glow of Mars. A smile began to grow on his face, and he turned around and walked up the steps, and sat in what was now "his" chair.

David gripped the chairs armrests, looked up, and banked the new ship with enormous power, in the direction of Mars.

Within seconds he had already sliced through the red planet's thin atmosphere. Dashing for the surface, he slowly leveled the ship to compensate for the approach of the incredible sight of Mariner Valley.

Chills of excitement rushed through him as he sank with tremendous speed into the valley's abyss. Rolling full circle, compensating with every maneuver, barely skimming over the surface of the mountain ranges deep within the valley's depths.

Banking hard right with magnificent momentum and speed, watching the valley's endless cliffs coming at him fast. Just before what seemed like a collision into the cliff walls, he banked the ship hard left, and remained completely sideways, slowly creeping closely up to their edge, paralleling them at inconceivable speed.

Rolling over full circle again, he headed back down into the deep crevasses and canyons, dodging and missing the skyscraper-like structures of unique, graceful splendor.

Just when the valley's end came rushing into view, he blasted with incredible velocity up and out of the great valley, exiting through the atmosphere, until the exquisite beauty of the heavens burst out before him.

He turned back to watch the Valley of the Mariners fade swiftly into the distance, and said, 'Until we meet again, new friend."

Turning slowly back to see the magnificent view of the Milky Way before him, he suddenly began to laugh out loud.

A very familiar voice had jumped into his mind, from a distance light years way.

It said, "You ride'em, Cowboy!"

****

www.ingramcontent.com/pod-product-compliance
Lightning Source LLC
Chambersburg PA
CBHW071239170526
45165CB00003B/1170